KODACHROME

KODACHROME

How Two Guys Named Leo Changed Photography

By Allan Weitz

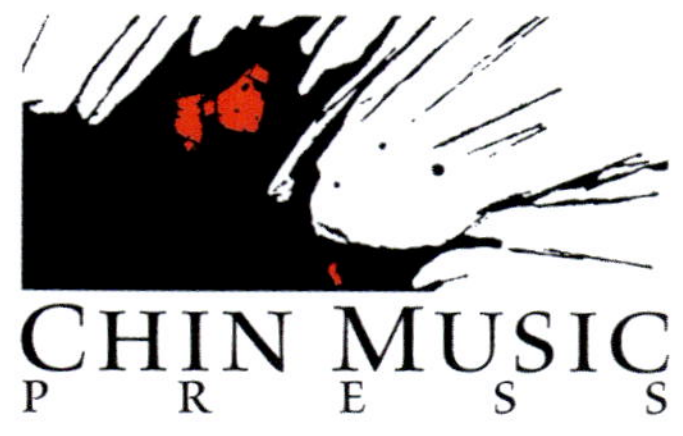

CHIN MUSIC PRESS

Printed Spring 2026

"There is no need any longer for us to pretend that the world is monochrome and to represent the glorious colored world in which we live by a gray ghost on a screen."

— Dr. Charles Edward Kenneth Mees, Kodak Research Center Director, introducing Kodachrome color transparency film on April 15, 1935.

A Chin Music Press original
Chin Music Press, Inc.
1501 Pike Pl # 329
Seattle, WA 98101
USA
www.chinmusicpress.com

Printed in South Korea
Library of Congress Control Number: 2026933039

CONTENTS

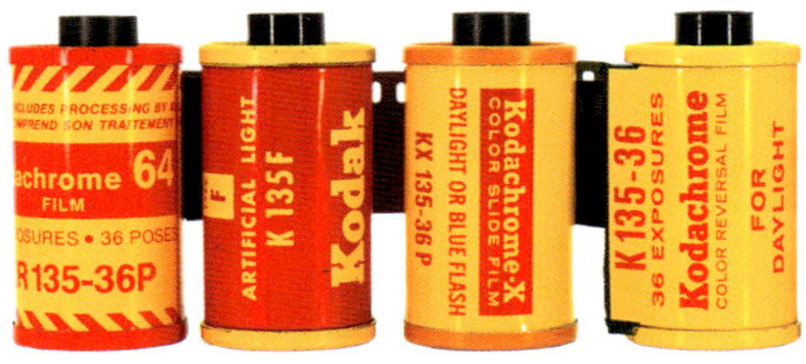

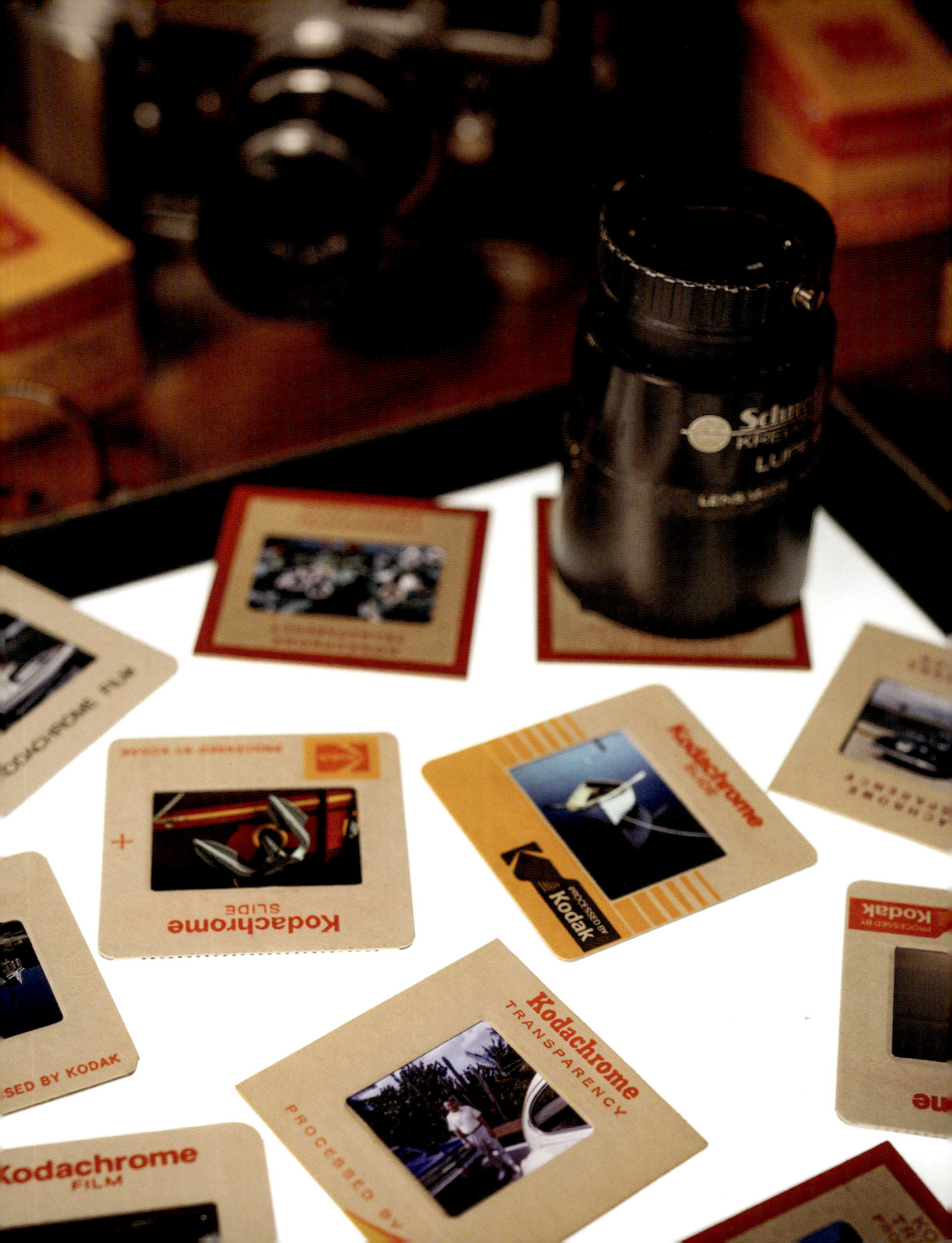
Kodachrome
TRANSPARENCY
Kodachrome
FILM
Kodachrome
SLIDE
PROCESSED BY KODAK
Kodak
PROCESSED BY Kodak
LUPE
LENS

Introduction

When Kodak announced it was discontinuing Kodachrome film, the news was akin to learning about the loss of an old friend, someone with whom I shared a long history. Kodachrome was my film of choice for close to three decades, as it was for many photographers, and soon it would be gone.

Kodachrome had a terrific run: seventy-four years, to be exact. What's interesting is how many advances were made in the field of camera technologies during Kodachrome's seven-decade run. At the time of its debut, cameras had three basic controls, none of which required batteries. There was a shutter-speed dial, aperture ring, and focusing barrel. That's it. Autofocus was still a good fifty years down the pike.

Thinking was mandatory when shooting Kodachrome. As tight as the film's grain structure was, that's how tight its exposure range was, especially under contrasty light. While you could afford to be "sloppy" when exposing negative film, with Kodachrome you only had about a third-of-a-stop wiggle room. Over or under a half-stop often made the difference between a dead-on exposure and a blown exposure. Any photographer worth his or her salt learned to "read" the light and shoot accordingly when using Kodachrome. The secret sauce consisted of a simple recipe: expose for the highlights and say a Hail Mary for the shadows.

Kodachrome was never a truly fast film. The original version had an ISO sensitivity of 10. Once the "faster" versions of Kodachrome came along, in 1961, most pros stuck to the ISO 25 and 64 versions, while keeping a few rolls of the 200-speed version tucked away "just in case." Not having the option of cranking up sensitivity levels to nose-bleed levels like we do today, Kodachrome taught one to navigate carefully when shooting in lower lighting conditions. A few independent labs advertised "pushed" processing for Kodachrome, but the results usually had the look of last-ditch attempts at covering up one's mistakes.

Hearing of Kodachrome's passing reminded me of a day long ago when my father and I stood on the rooftop of our apartment house, in Marine Park, Brooklyn. In the distance we watched as the first commercial jetliner, a Boeing 707, took its first scheduled lift-off from JFK, or Idlewild Airport as it was known back in 1958.

For my father, who was a master mechanic who could dismantle, repair, and rebuild anything that could fly during World War II, the event was particularly poignant. As we watched the gleaming jet bank over our home, I remember him say wistfully to no one in particular, "They finally got the kinks out of piston engines, and now...?"

The last time I shot film was August 2000. Fittingly, it was photographing a couple of antique speedboats. And though I learned the art and craft of photography with a film camera, I've never looked back. Is digital perfect? Nope, and maybe it never will be. But just as jets have relegated piston-powered aircraft to the Cessna crowd and regional puddle-jumpers, film cameras are proving to be quite popular among young, post-analog creative types.

Just as we'll never see another Lockheed Constellation taxi down the runway at JFK, so goes Kodachrome. But then again, no aircraft can ever lay claim to having a best-selling record, a national park, or a motion picture starring Ed Harris named after it.

— Allan Weitz

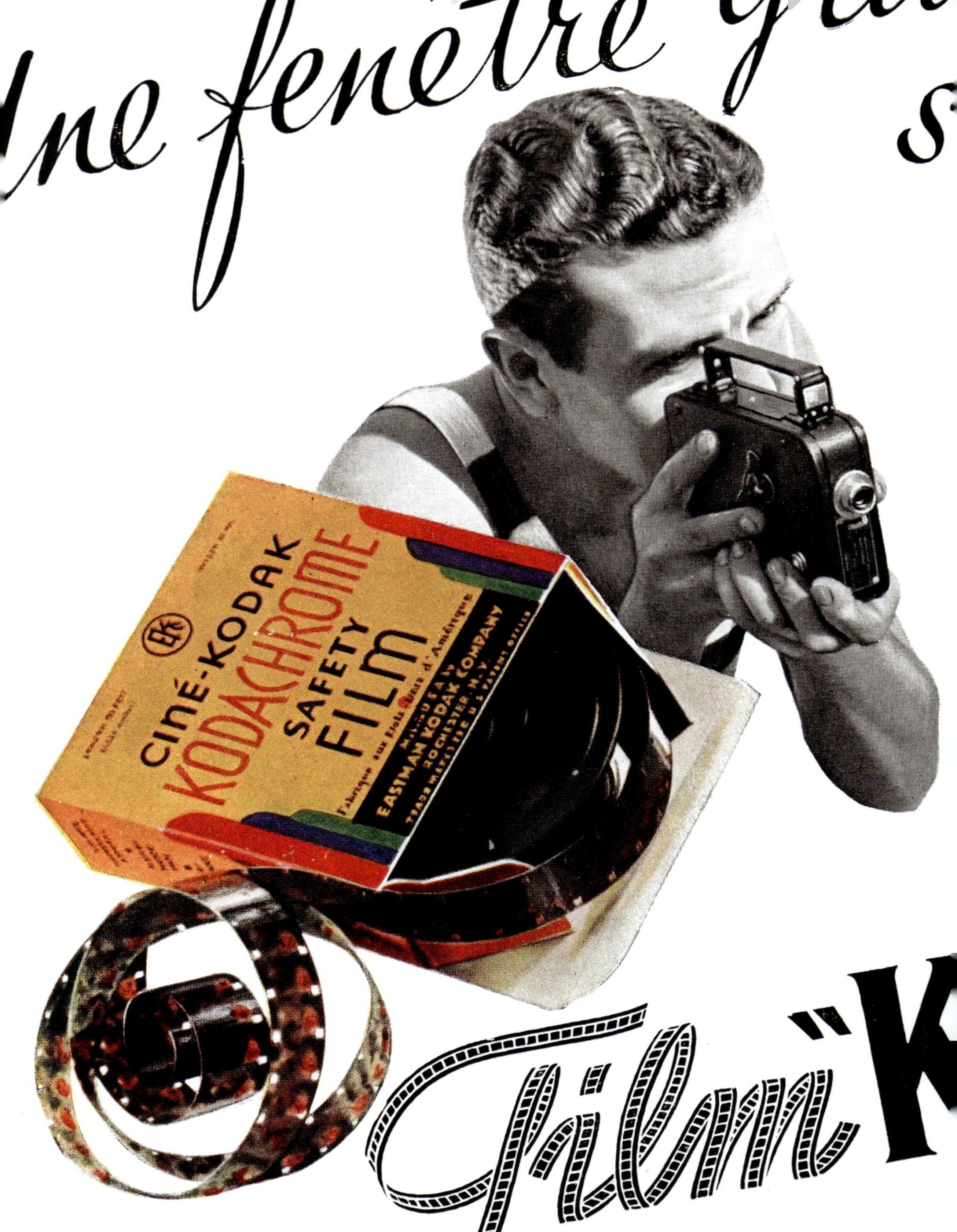
Une fenêtre qui s
CINÉ-KODAK
KODACHROME
SAFETY
FILM
EASTMAN KODAK COMPANY
ROCHESTER N.Y.
Fabriqué aux États-Unis d'Amérique
Film "K

TRUE COLORS

The Origin of Kodachrome Film

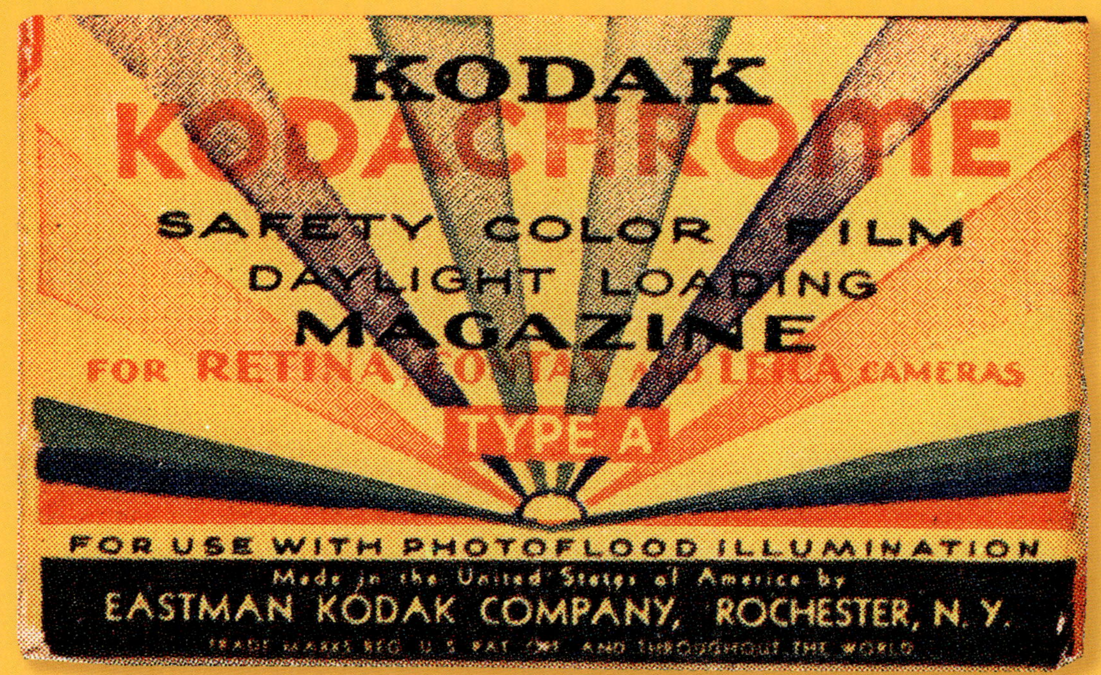

KODAK
KODACHROME
SAFETY COLOR FILM
DAYLIGHT LOADING
MAGAZINE
FOR RETINA AND PAX AND LEICA CAMERAS
TYPE A
FOR USE WITH PHOTOFLOOD ILLUMINATION
Made in the United States of America by
EASTMAN KODAK COMPANY, ROCHESTER, N. Y.
TRADE MARKS REG. U.S. PAT. OFF. AND THROUGHOUT THE WORLD

The Birth of Kodachrome

On April 15th, 1935, Eastman Kodak introduced Kodachrome, a daylight-balanced color transparency film capable of reproducing colors that were far brighter and more natural-looking than existing color films. The film, which was available for 8mm and 16mm motion picture cameras, was called Kodachrome A. It was daylight balanced and had a light sensitivity of ISO 10. In 1936 35mm and 828 cartridge versions of Kodachrome were introduced for still cameras.

For the next seventy-four years, successive generations of Kodachrome would continue to cast shade on competitive color transparency films. Kodachrome would most likely have continued in its role as the finest color transparency film to this very day if not for a newer Kodak technology introduced in the late 1990s: digital imaging, which in the space of a decade would all but sound the death knell of analog photography.

The seeds of Kodachrome were planted in 1916, when Leopold Mannes and Leopold Godowsky Jr.—two sixteen-year-old classically trained musicians with keen interests in photography—stepped out of a movie theater after seeing a short color feature called "Our Navy."

The movie was filmed using a new color film process called Prizma Color, which was touted as having better color than other film processes of the day. While there's no record of what Mannes and Godowsky thought of the costumes or story line, one thing they did agree upon is that the highly touted Prizma Color process was underwhelming. To their eyes, the colors were as muted and dim as any of the existing color film processes of the day.

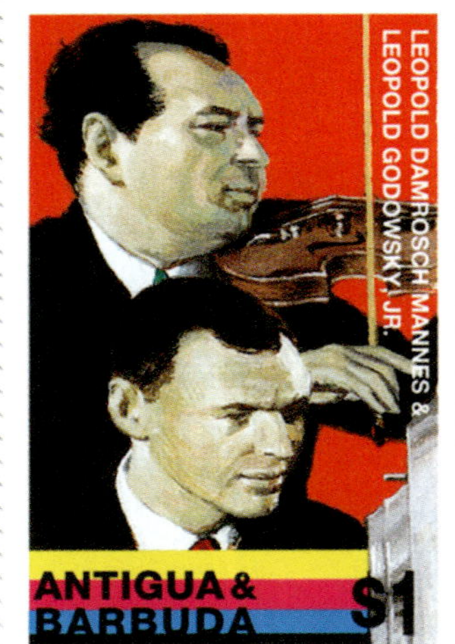

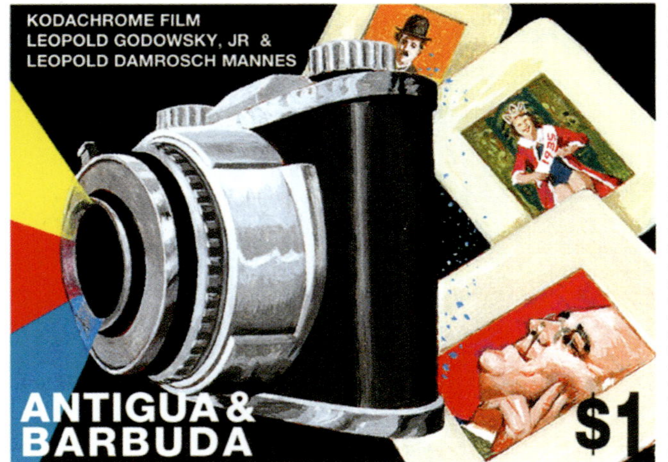

1998 postage stamps from Antigua & Barbuda honoring Leopold Mannes and Leopold Godowsky.

The two Leopolds figured that if the major players could not develop a motion picture process that would faithfully reproduce color, maybe they should take a crack at it themselves. As Godowsky was quoted as saying many years later, "We were blissfully ignorant." Although it took close to two decades and an unimaginable amount of labor, capital, and disappointment before the fruits of their labor came to market, the "two Leos," or "God and Man" as they came to be known among their peers, managed to create a color film that would become the gold standard of color photography.

Leopold Damrosch Mannes was born on December 26, 1899. His maternal grandfather, Leopold Damrosch, was a world-famous conductor. His parents, David and Clara Mannes, founded the Mannes College of Music, in Manhattan. Mannes studied piano and physics at Harvard and went on to earn a Pulitzer Music Scholarship and a Guggenheim fellowship to study music in Italy.

Leopold Godowsky Jr. was born on May 27, 1900. The son of composer and pianist Leopold Godowsky, the younger Godowsky studied violin at UCLA and went on to become a soloist and first violinist with the San Francisco

Symphony and Los Angeles Philharmonic Orchestras. His wife Frances was George and Ira Gershwin's sister. They had a son named Leopold Godowsky III, who went on to become a celebrated concert pianist in his own right. Godowsky dropped the 'Jr' designation from his name shortly after his father passed away in 1938.

Most of their earlier experiments took place in the kitchens and bathrooms of their parents' apartments on Manhattan's Upper West Side. Processing times and chemical temperatures were critical and had to be performed in total darkness. Not trusting the accuracy of commercial timers, they would whistle the last movement of Brahm's C-minor symphony to the rhythm of a metronome set at two beats per second.

When informed they had overstayed their welcomes at their respective parents' homes, they continued their work after hours in rented hotel rooms.

Taking advantage of family connections within the New York theater world, they managed to gain access to the projection booth at S.L. 'Roxy' Rothafel's state-of-the-art Rialto movie theater in which they could view the results of their efforts in their intended environs.

After a number of false starts, they formulated a promising film emulsion that caught the attention of Kodak's chief scientist, Dr C.E. Kenneth Mees. In 1912 Mees was hired by Kodak's founder, George Eastman, to design and staff what would ultimately become the world-renowned Kodak Research Laboratory, located in Kodak Park in Rochester, New York. Mees was a mere thirty years old at the time.

Mees understood the technical difficulties Mannes and Godowsky were facing formulating the color couplers and preventing color migration between layers. Confident in their abilities, Mees had color-sensitive, multiple-layer emulsions prepared to Mannes and Godowsky's specifications and sold these materials to them at cost to help extend their financial resources and speed up their progress.

As a result of a chance conversation with a fellow passenger on a trans-Atlantic crossing in the Fall of 1922, a meeting was arranged between Mannes, Godowsky, and Lewis Strauss, who was a junior associate at Kuhn, Loeb & Partners, which was a well-respected Manhattan investment firm. Strauss was impressed enough to offer them a twenty thousand dollar loan to further their research. The two Leopolds had their first investors.

They opened their first laboratory in 1924 and, by 1931, they secured forty patents. Impressed with their headway, Mees offered Mannes and Godowsky an opportunity to continue their research at Kodak's new Research Laboratory. In 1930, all parties came to an agreeable arrangement. Mannes and Godowsky would be paid thirty thousand dollars up front, much of which went towards paying off their remaining research loan to Kuhn, Loeb, and Strauss. Mannes and Godowsky also agreed to salaries of $7,500 per year, and royalties on all patents filed prior to signing on with Kodak.

Interestingly, in 1922, after being forewarned about their impending expulsion from their respective parents' apartments, Mannes and Godowsky managed to get an appointment to meet with George Eastman during one of his trips from Rochester to Manhattan. Their goal was to gain financial backing from him in order to continue

Postcard showing the Kodak Park complex in Rochester, NY.

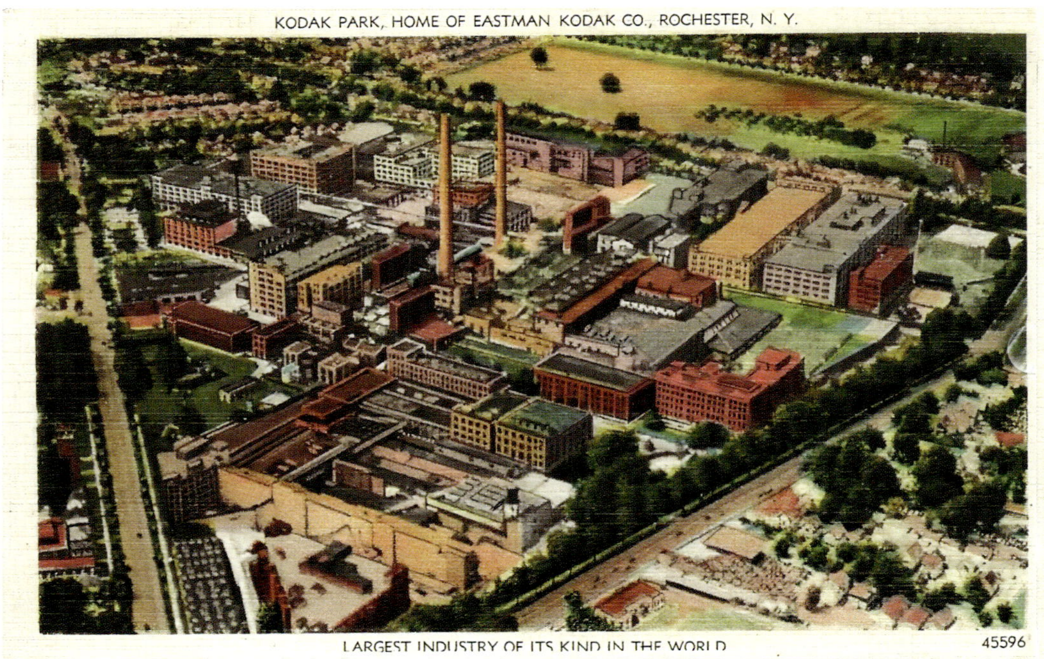

their work. And while he displayed interest in their accomplishments and promised to see what he can do to help them along, he never did get back to them. Fortunately, it wasn't long before they secured their twenty thousand dollar loan from Kuhn, Loeb & Partners.

After considerable false starts and setbacks, Kodak introduced Kodachrome slide film in April 1935. Little did they know at the time it would turn out to be a film that literally changed the way consumers captured and saved special moments in time.

Kodachrome was by no means the first color film. This honor belongs to a Scottish chemist named James Clerk Maxwell who, seventy-five years earlier (1861), demonstrated a three-color additive color process that involved photographing a scene using three individual black and white coated plates, each exposed through a blue, green, or red filter.

These plates would then be developed. Once dry, a positive slide was made from each of the three negatives. The three positive slides would be projected using three perfectly aligned projectors that were fitted with the same blue, green, and red filters through which the negatives were originally exposed. It was a labor-intensive undertaking, but it worked. The projected image was a fairly accurate color rendition of the original scene.

Magical as it was, Maxwell's color process, along with the other additive processes that followed (Lumier, Autochrome, Finlaycolor, Dufaycolor, and Agfacolor), produced imagery with muted colors that tended to be dim and pale, which wasn't surprising considering the emulsions absorbed up to seventy percent of the light passing through them. While there was an element of novelty to early color films,

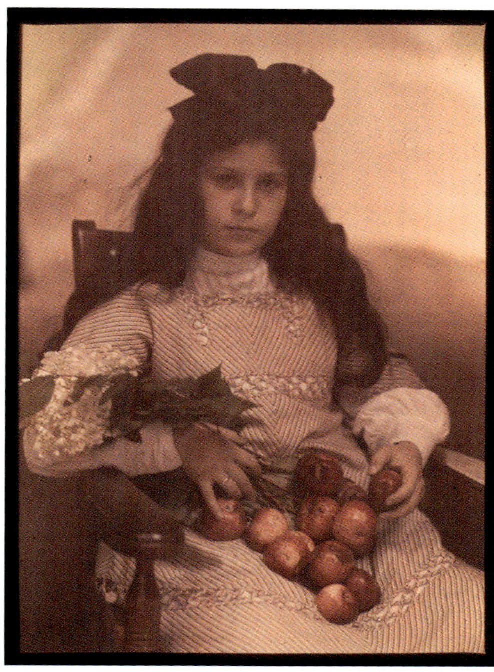

An Autochrome color photograph by Alfred Stieglitz circa 1907 (Art Institute of Chicago).

the colors lacked "snap, crackle, and pop." Kodachrome changed all that.

Unlike its predecessors, Kodachrome was a subtractive system that exposed a single multi-layered (cyan, magenta, and yellow) film stock. Though still relatively slow (ISO 10), Kodachrome was the "every man film," meaning anybody could capture colorful home movies with small, easy-to-use cameras that Kodak was producing to fit every budget. The color quality of Kodachrome proved to be bright and colorful enough to relegate most existing color processes to the history bins within a decade.

The earliest consumer versions of Kodachrome introduced in 1935 involved the use of what is called a "dye destruction process" that required several hours to process including three separate processing baths with a dry cycle between each bath and monitoring at each step with a microscope. The process worked but proved to be far too complex to sustain on a long-term basis.

In 1936 Dr. Mees hired Dr Wesley T. (Bunny) Hanson Jr., age twenty-three, to figure out a way to simplify and streamline the process in a bid to make Kodachrome both practical and profitable to manufacture and market to the public. Two years later an updated version of Kodachrome as introduced that replaced the complex and time consuming dye destruction process with a process that re-exposed the film to colored light while continuing to incorporate color couplers in the processing solutions.

The same structural system of using three stacked light-sensitive layers recording blue, green, and red respectively was used in all of Kodak's subsequent Kodachrome film variations and development processes, all the way through Process K-14.

It should be noted the name Kodachrome was first used about twenty years earlier—coincidently around the same time that the two Leos began mixing chemistry in the bathrooms of their parents' apartments. The original Kodachrome was a complex two-color (blue-green, red-orange) subtractive process developed by a Kodak chemist named John Capstaff. While the film's color palette was well suited for portraiture, it proved to be dismal for landscapes due to its lack of a blue layer. The advent of World War I ultimately put the kibosh on any further development of Capstaff's color film process.

It's also worth noting Mannes and Godowsky gave considerable credit to the research of Rudolph Fischer, a German chemist who in 1911 invented color couplers and the processing procedures that were central to Mannes and Godowsky's efforts in perfecting the Kodachrome color process.

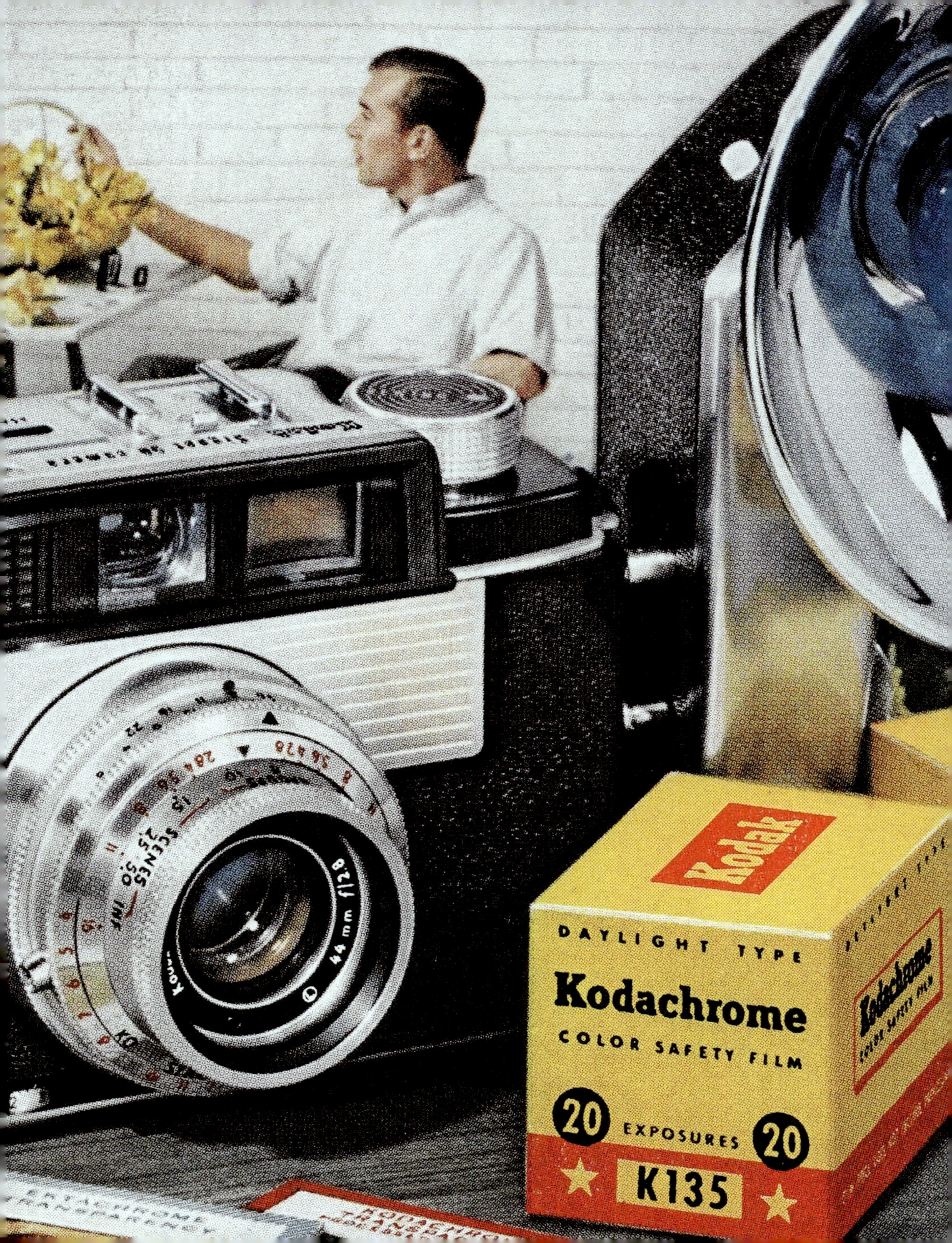
Kodak
DAYLIGHT TYPE
Kodachrome
COLOR SAFETY FILM
20 EXPOSURES 20
K135

KODACHROME EVOLVES

A Full Line of Products & Services

The Original: Kodachrome A

The first rolls of Kodachrome A introduced in 1935 eschewed bright, beautiful colors unlike anything photography enthusiasts of the day had ever seen before. The film was slow (ISO 10) and was balanced for daylight use.

Initially available for 8mm and 16mm motion picture cameras, 8-exposure 35mm and 828 cartridges were made available a year later for Kodak Bantam and comparable 828-format cameras from other manufacturers.

Bantam 828 films, a format which has been long discontinued, was essentially 35mm minus the sprocket holes, which allowed for about thirty percent additional image area (40 x 28mm vs. 24 x 36mm) compared to 35mm film.

The original formulations of Kodachrome and its processing procedure proved difficult to maintain in terms of color consistency. As a result, a reformulated Kodachrome development process was developed that was less complex and delivered consistently reliable results. Over the next seventy years there would be two additional reformulations of Kodachrome and processing procedures, each a notable improvement over its predecessor.

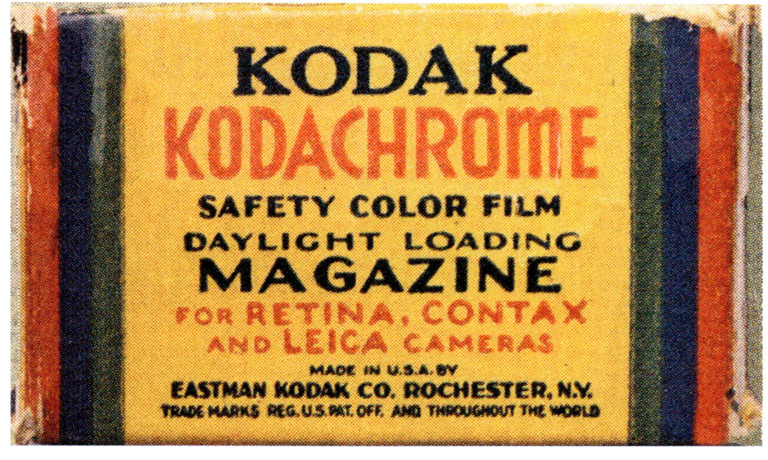

KODAK
KODACHROME
SAFETY COLOR FILM
DAYLIGHT LOADING
MAGAZINE
FOR RETINA, CONTAX
AND LEICA CAMERAS
MADE IN U.S.A. BY
EASTMAN KODAK CO. ROCHESTER, N.Y.
TRADE MARKS REG. U.S. PAT. OFF. AND THROUGHOUT THE WORLD

Kodachrome II & Kodachrome-X

Although photographers and movie enthusiasts loved the original, the sluggish ISO 10 sensitivity of Kodachrome A made it difficult to use under all but the brightest of lighting conditions.

With an ISO of 25, Kodachrome II, which was introduced in 1961, was two and a half stops faster, finer grained, sharper, and had better color rendition than the original Kodachrome emulsion. Kodachrome II had a dynamic range of about eight stops (3.6-3.8D), or about 6.3 shades of gray. A year later (1962), Kodachrome-X was introduced, sporting slightly bolder colors than Kodachrome II, along with an ISO of 64.

Though slow by today's standards, having a Kodachrome emulsion with a "blistering" ISO of 64 turned low-light photographers into happy campers. In addition to added speed, Kodachrome II and Kodachrome-X gained the advantages of Kodak's new K-12 film processing formulation, which replaced the previous K-11 process. An indoor version of this Kodachrome formula (3400K, ISO 40) was also made available at this time.

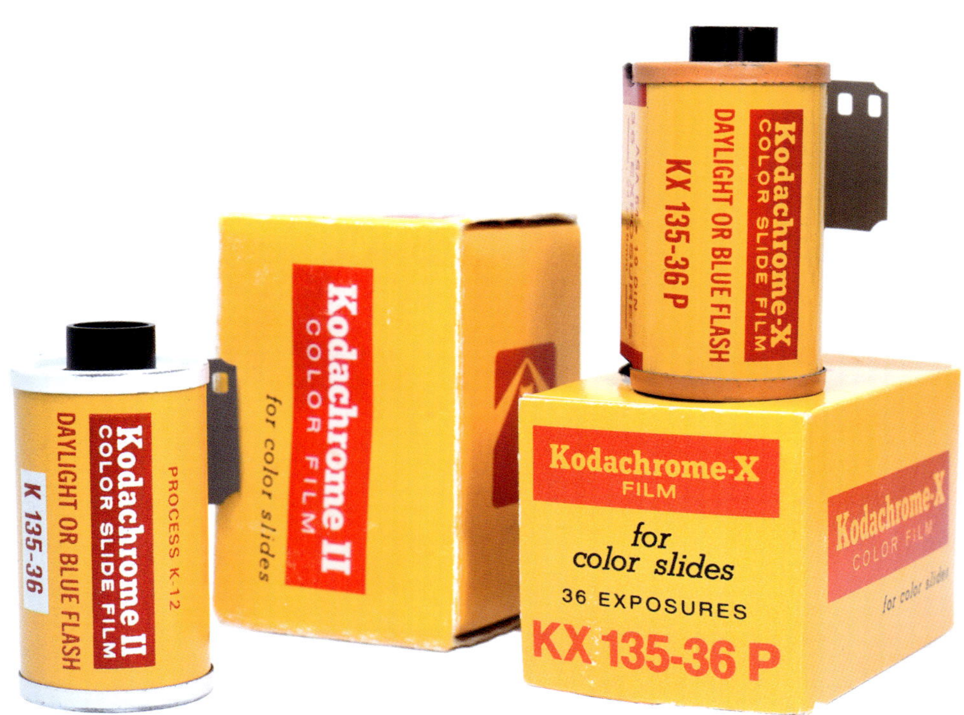
Kodachrome II
COLOR SLIDE FILM
DAYLIGHT OR BLUE FLASH
PROCESS K-12
K 135-36
Kodachrome II
COLOR FILM
for color slides
Kodachrome-X
COLOR SLIDE FILM
DAYLIGHT OR BLUE FLASH
KX 135-36 P
Kodachrome-X
FILM
for color slides
36 EXPOSURES
KX 135-36 P
Kodachrome-X
COLOR FILM
for color slides

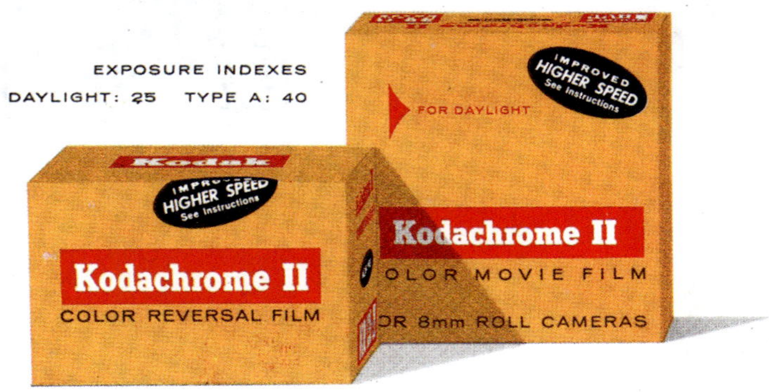

Faster and better…
The facts about Kodachrome II Film

Speed: 2½ times as fast as regular Kodachrome Film. Exposure indexes: Daylight, 25; Type A, 40. You can stop down your lens for greater depth of field. Shoot in less light or at faster shutter speeds. All with a definite improvement in picture quality! *Sharpness* is increased, grain even further reduced to record even finer detail—particularly important to 8mm users. *Color rendition* is improved. Bright colors—reds, greens, yellows—are more vivid, pastels cleaner. *Contrast* is lower. Slides and movies reproduce a greater range of brightness with more detail in highlights and shadows. *Exposure latitude* is wider. It's hard *not* to get a pleasing rendition of the color you see. *Available* in 135, 828, 8mm and 16mm sizes,* in limited quantities. Larger quantities will be coming. Priced slightly higher than regular Kodachrome Film. If your dealer is out of stock, please ask for it again.

EASTMAN KODAK COMPANY, Rochester 4, N. Y.

*For still cameras Kodachrome II Film is available in Daylight Type only. It can also be used indoors with blue flashbulbs or electronic flash. Both Daylight and Type A are available for movie cameras.

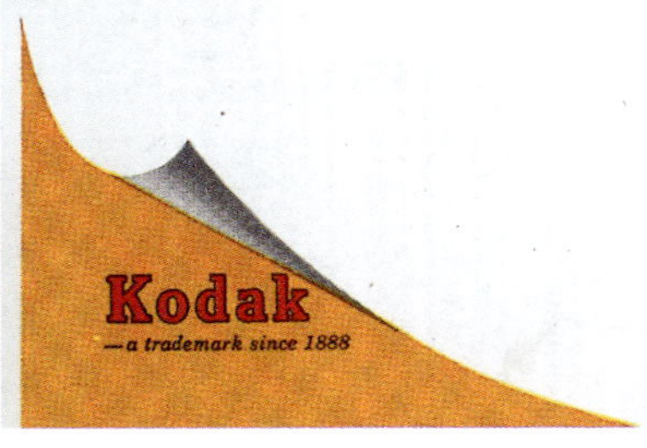

1961, USA

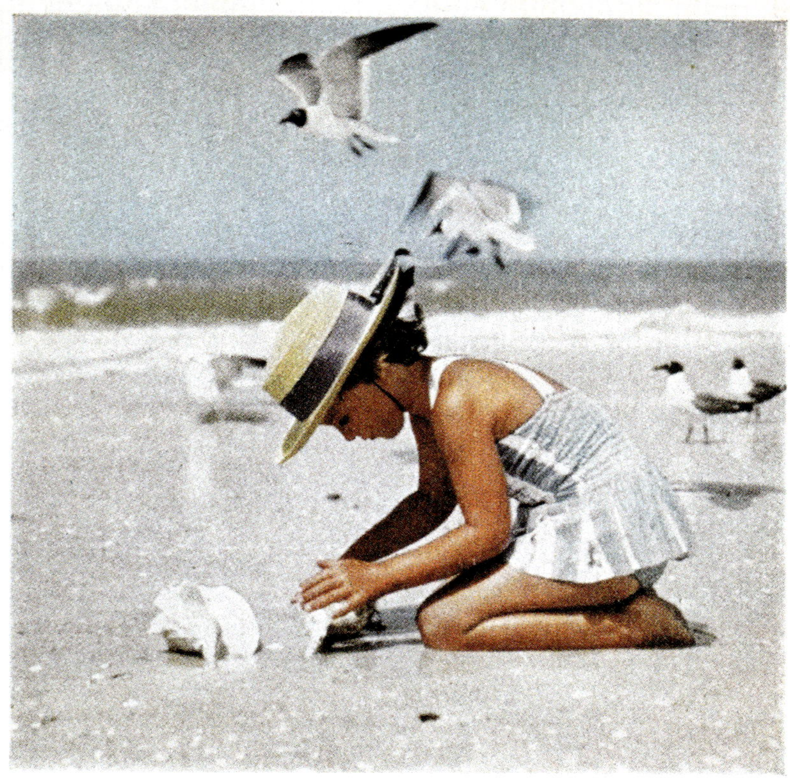

Click.

'There are moments in life
worth purchasing with worlds.'

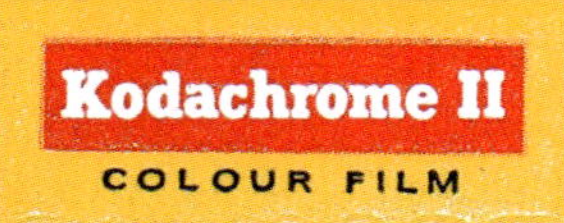

Kodachrome II
COLOUR FILM

When you click a camera, you stop an instant of Time. It seems a fairly
cosmic thing to do. But it's incredibly simple. All you need's a camera, Kodak
colour film, and the wit to realize that a moment only happens once.

'Kodachrome' is a registered trade mark of Kodak Ltd. The line we quote comes from Fielding's Amelia.

♠ 1963, UK

Kodachrome 25 & Kodachrome 64

Kodachrome 25 and Kodachrome 64, which were introduced along with Kodak's improved K-14 processing formula, were the last of the Kodachrome upgrades. Introduced in 1974, Kodachrome 25 and Kodachrome 64 were available in 35mm, 110, 126, and 120 formats and had dynamic ranges of up to twelve stops.

Just as Kodachrome II and Kodachrome-X were notable improvements over the original Kodachrome transparency film line, Kodachrome 25 and Kodachrome 64 were comparably sharper and featured improved color and contrast, compared to earlier versions of Kodachrome.

In 1983 Kodak introduced Kodachrome Type A (PKA), an indoor ISO 40 film, Kodachrome 25 Professional (PKM), and Kodachrome 64 Professional (PKR) for the professional market. These emulsions were manufactured to stricter technical parameters and required refrigeration.

Production of the 35mm versions of Kodachrome 25 and Kodachrome 64 were discontinued in 2009. The 110, 126, and 120 medium format versions were discontinued even earlier.

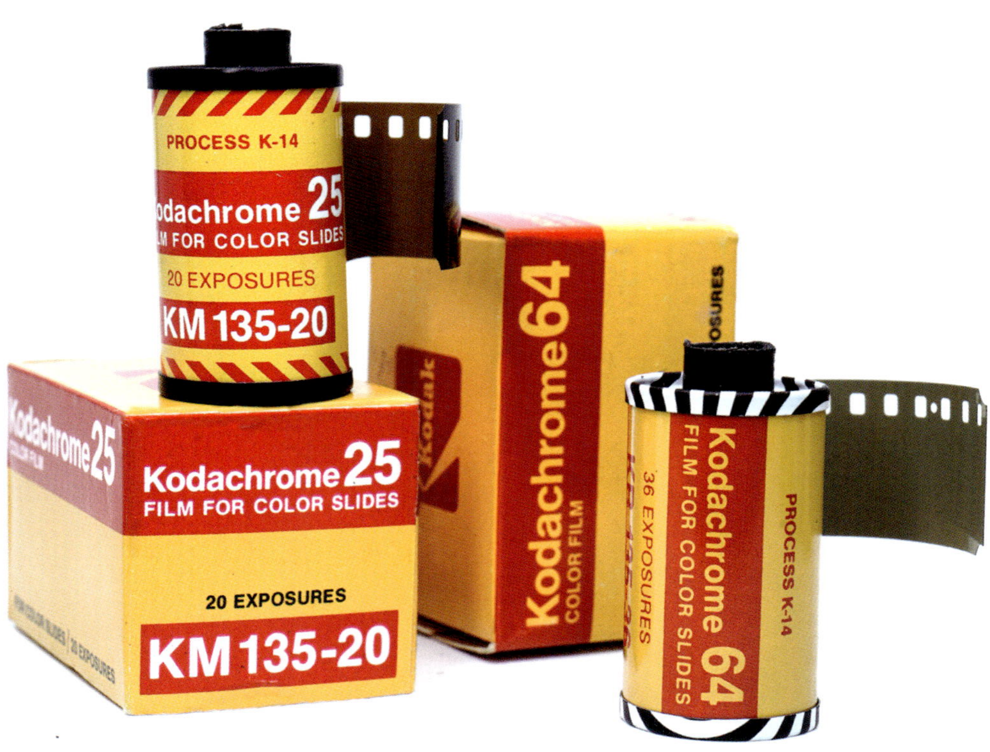

PROCESS K-14
Kodachrome 25
FILM FOR COLOR SLIDES
20 EXPOSURES
KM 135-20
Kodachrome 25
FILM FOR COLOR SLIDES
20 EXPOSURES
KM 135-20
Kodachrome 64
COLOR FILM
Kodachrome 64
FILM FOR COLOR SLIDES
PROCESS K-14
36 EXPOSURES

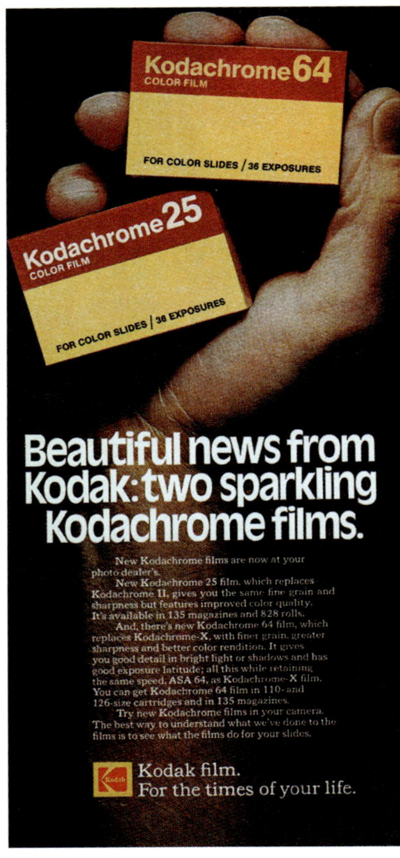

1974, USA ◆

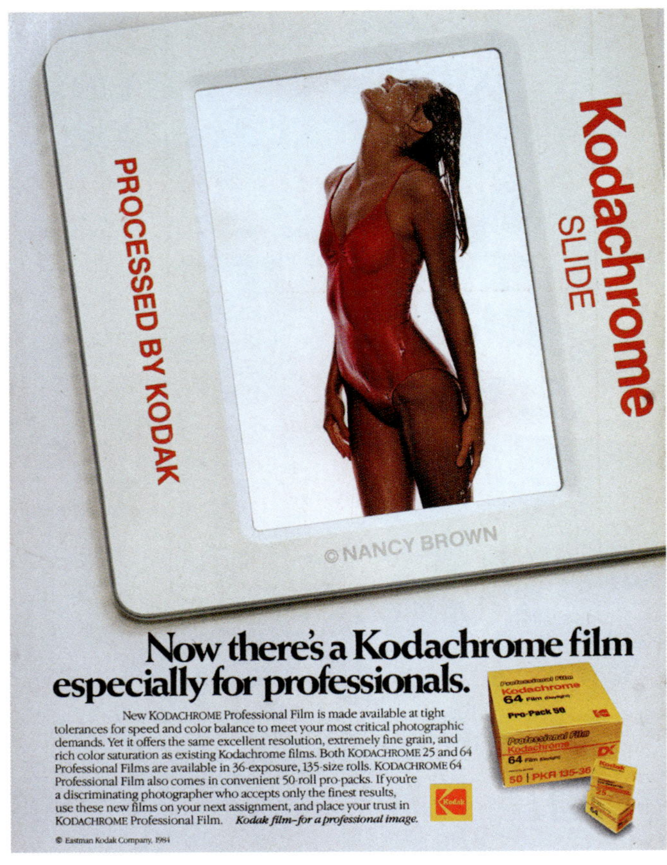

1984, USA ◆

1985, USA ▶

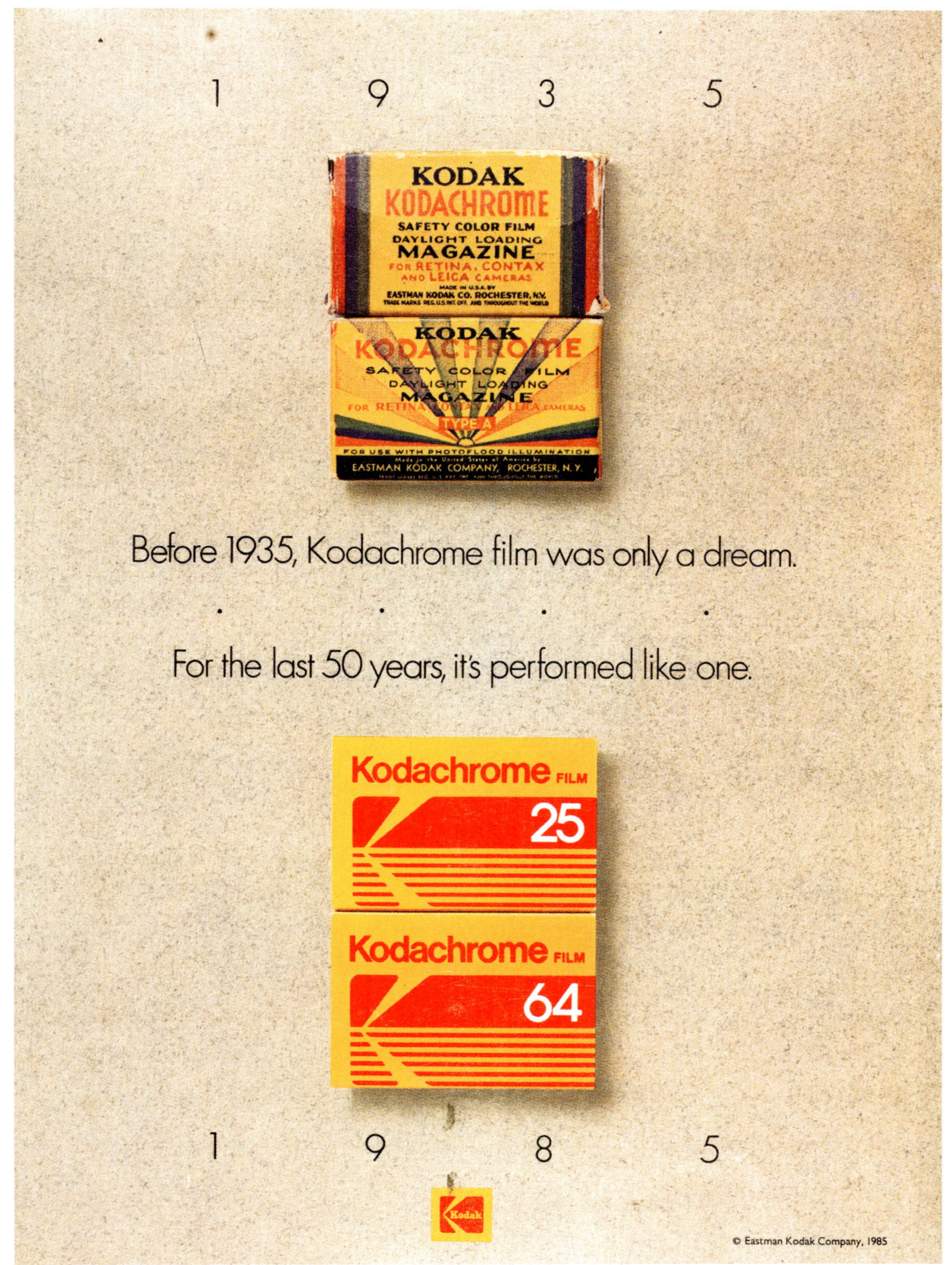

1 9 3 5
KODAK
KODACHROME
SAFETY COLOR FILM
DAYLIGHT LOADING
MAGAZINE
FOR RETINA, CONTAX
AND LEICA CAMERAS
MADE IN U.S.A. BY
EASTMAN KODAK CO. ROCHESTER, N.Y.
TRADE MARKS REG. U.S. PAT. OFF. AND THROUGHOUT THE WORLD
KODAK
KODACHROME
SAFETY COLOR FILM
DAYLIGHT LOADING
MAGAZINE
FOR RETINA, CONTAX AND LEICA CAMERAS
TYPE A
FOR USE WITH PHOTOFLOOD ILLUMINATION
Made in the United States of America by
EASTMAN KODAK COMPANY, ROCHESTER, N.Y.
TRADE MARKS REG. U.S. PAT. OFF. AND THROUGHOUT THE WORLD
Before 1935, Kodachrome film was only a dream.
For the last 50 years, it's performed like one.
Kodachrome FILM 25
Kodachrome FILM 64
1 9 8 5
Kodak
© Eastman Kodak Company, 1985

Kodachrome 200

Before the days of the seven-digit ISO sensitivities, Kodak's introduction of Kodachrome 200, which was about a stop-and-a-half faster than Kodachrome 64, was considered a big deal among sports photographers and others who worked in less than optimal lighting conditions.

Kodachrome 200 was introduced in 1986. The faster film, daylight balanced with a dynamic range of approximately eight stops (2.3D), was considered an absolute godsend by photographers who preferred the color rendition of Kodachrome over Ektachrome and comparable E-6 reversal films. Though somewhat grainier and more contrasty than Kodachrome 25 and Kodachrome 64, Kodachrome 200 was significantly finer-grained when compared to other high-ISO film stocks.

During the Olympic Games of the late eighties and early nineties, Kodak sponsored the photographic portion of the Olympic Press Center. In addition to free Kodak film and processing, photographers were able to push Kodachrome 200 to ISO 1500, which greatly expanded the imaging opportunities of the attending photographers. As a bonus, the magenta cast that resulted from pushing the film speed neutralized the green cast of the stadiums mercury halogen lamps, making it a win-win situation for the photographers and the print technicians who had to color-separate these photographs for publication.

Rapid improvements in digital technologies, paired with an equally rapid decline in film sales, led to the discontinuation of Kodachrome 200 in 2008.

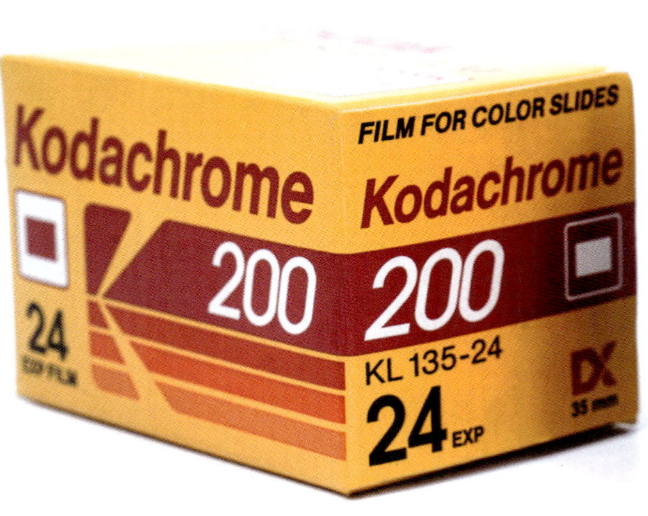
FILM FOR COLOR SLIDES
Kodachrome
Kodachrome
200
200
KL 135-24
24
EXP
DX
35 mm
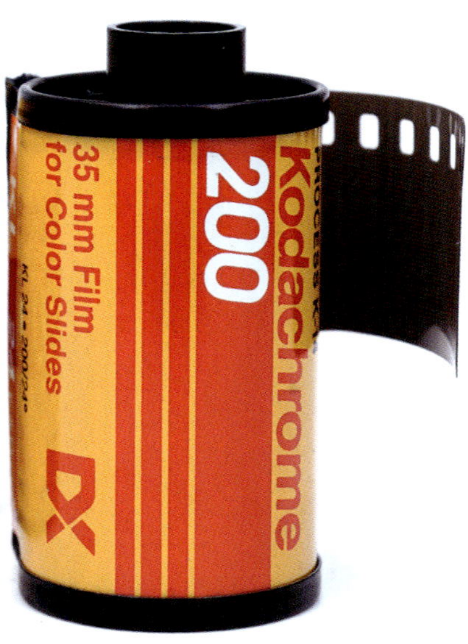
35 mm Film
for Color Slides
Kodachrome
200
DX

Nouveau film Koda

Bien qu'il soit nette

sa ressemblance a

ne vous aura pas é

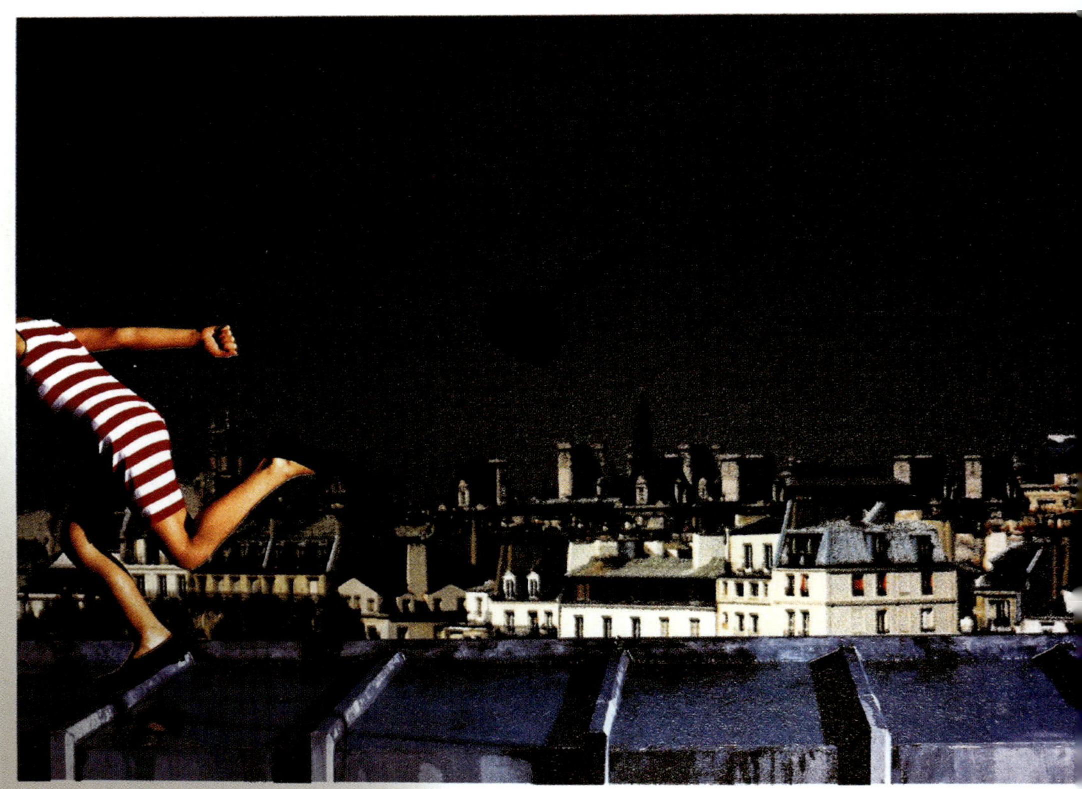

1988, France ⬥

chrome 200.
ment plus rapide,
vec le Kodachrome
happé.

Pousser la perfection plus loin, s'ouvrir sans cesse de nouveaux horizons : c'est l'objectif de tout créateur, et celui de KODAK en lançant le nouveau film Kodachrome 200. Ce film aux caractéristiques exceptionnelles vient compléter une gamme désormais légendaire : celle des films Kodachrome 25 et 64 ISO. Rendu chromatique d'une vérité absolue, haute saturation des couleurs, finesse de l'image, piqué inouï, excellente définition dans les ombres, il a toute la magie du Kodachrome. Celle qui en fait le film de prédilection des grands professionnels.

Avec 200 ISO, il ne fuit devant aucune difficulté. Vous pouvez désormais travailler en faible lumière, saisir les actions rapides, augmenter la profondeur de champ ou utiliser un téléobjectif sans perdre en beauté ni en résolution. Ces performances, impossibles jusqu'alors dans les hautes sensibilités, ont pu être réalisées grâce à la technologie de pointe des grains T. Appliquée pour la première fois à un film Kodachrome, elle permet de concilier rapidité, absence de grain et très haute définition. Le Kodachrome était la référence dans les basses sensibilités, il l'est désormais en toutes circonstances.

Conçu pour être utilisé en lumière du jour, le film Kodachrome 200 accepte bien entendu le flash électronique sans filtre. Comme tous les films de haute qualité, il est recommandé de le faire développer le plus rapidement possible après exposition afin de ne pas altérer l'image latente. Il permet par ailleurs toutes les techniques d'impression, inversible direct, avec internégatif ou procédé de dye-transfert.

Un seul film pouvait rivaliser avec le Kodachrome : un autre Kodachrome. D'une grande rapidité et d'une transparence inégalée, il vous permet de vous dépasser dans toutes les situations.

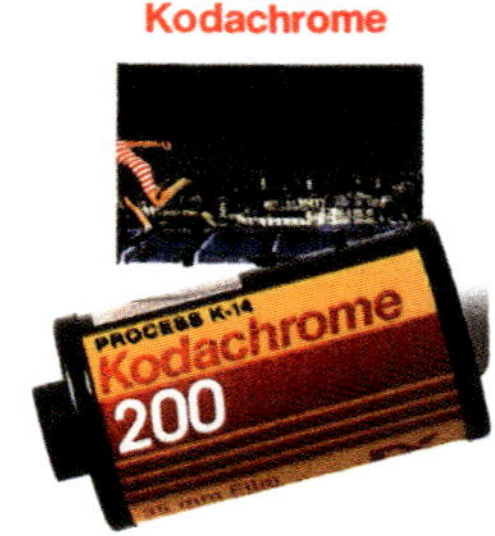

KODAK. **TOUJOURS UN DÉCLIC D'AVANCE.**

Project 237

In 1985, Kodak undertook a program to expand the Kodachrome lineup to better match the speed choices of Ektachrome films. Fearing leaks to competitors, Kodak gave the project a special program "P" name and number, which in this case was "P-237."

The goal was to reformulate Kodachrome 25 along with new 100-speed, 200-speed and 400-speed Kodachrome variations. The new films would be manufactured using new reversal color film coating machines. They would also use Kodak's latest silver halide T-grain technologies, which improved grain and sharpness while reducing silver consumption.

Alas, all did not go as planned. For starters, the new T-grain technologies failed to improve the performance of Kodachrome 25 significantly, forcing Kodak to scale back and settle for modest improvements minus the T-grain technology.

Kodachrome 100 was terminated after it was assessed as being too close in speed to Kodak's extremely popular (and extremely profitable) Kodachrome 64. In turn, Kodachrome 64 received a few image-enhancing tweaks similar to those implemented for Kodachrome 25.

Hopes for a 400-speed Kodachrome were abandoned when Kodak's film engineers failed to achieve acceptable grain positions for speed, sharpness, and pressure performance in all three color film layers. The new film simply could not compete against any of the existing 400-speed reversal films.

To add insult to injury, the newly developed reversal color film coating machines were taken offline, and all film manufacturing was transferred to a different coating facility, requiring reformulations to the new films. Kodachrome 64 and 200 proved to be adaptable, but Kodachrome 25 failed to produce acceptable results.

In all, the only new film to emerge from P-237 was Kodachrome 200, which, while not as sharp or vibrant as its slower siblings, was nonetheless a welcome higher-ISO addition to the Kodachrome family.

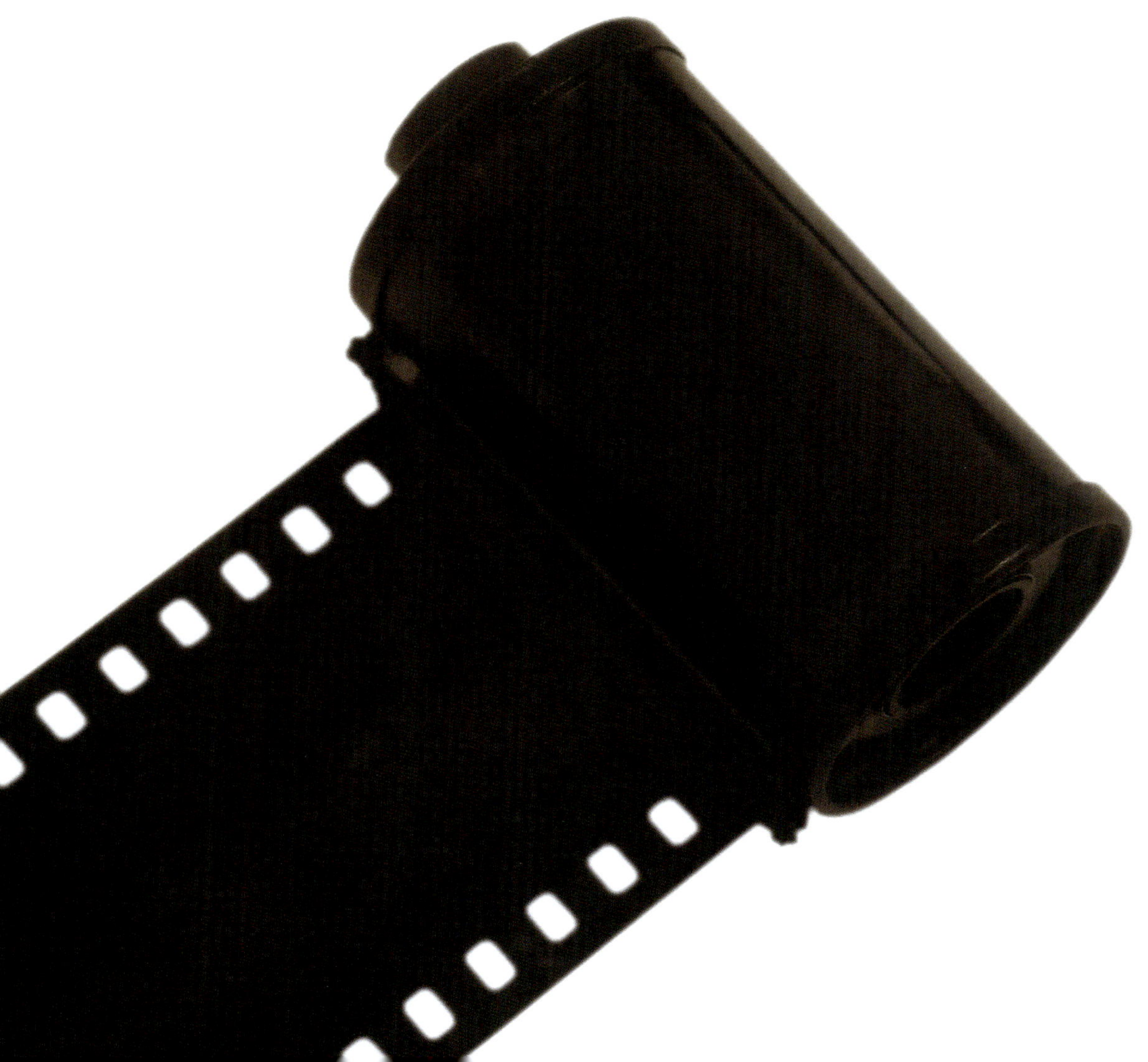

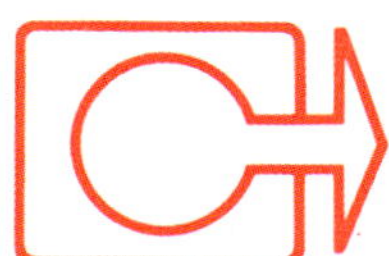

Kodak brings the <u>Instamatic</u> Camera idea to movies!

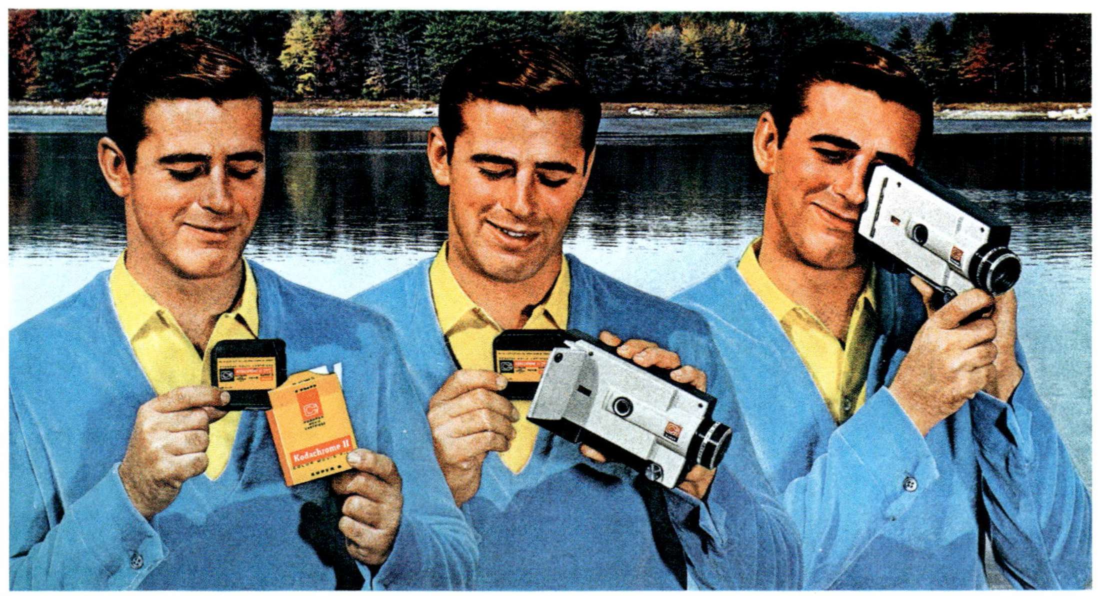

Open…   drop in…   shoot!

Kodak announces <u>Instamatic</u> Movie Cameras—and a complete new system of movie-making. A new kind of movie camera loads instantly, automatically. A new kind of movie film gives you clearer, sharper movies than ever before. Just slip the <u>Kodapak</u> Movie Cartridge into the camera and shoot a complete roll of brilliant color movies. No threading. No winding. No need to flip the film! In fact, your hands never even touch the film! And on your screen—see movies that are more than ever *the nearest thing to life itself!*

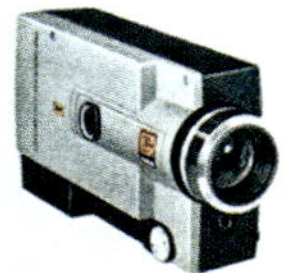

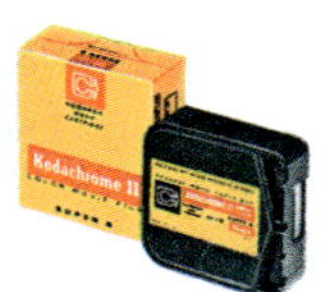

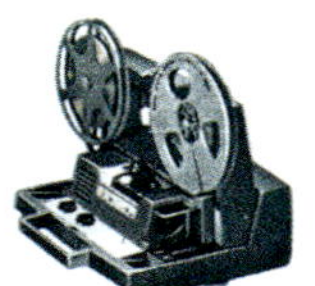

Kodak has redesigned the movie camera. New KODAK INSTAMATIC Movie Cameras load instantly. No threading. An electric motor drives the film for you automatically. M2 model, less than $50; M4 model with electric eye, less than $75; M6 zoom model (above), less than $175.

Kodak has redesigned the movie film with 50% more picture area for better, sharper movies. The new KODAPAK Movie Cartridge is factory-loaded with improved KODACHROME II Film in the new Super 8 format—giving you 50 feet of uninterrupted shooting.

Kodak has redesigned the movie projector. The KODAK INSTAMATIC M70 Movie Projector shows any scene at normal, fast, or slow-motion speeds—forward and reverse. "Still" projection, too. Brilliant illumination, automatic threading. Less than $160. Other models from less than $65.

EASTMAN KODAK COMPANY, Rochester, N.Y.

Prices subject to change without notice.

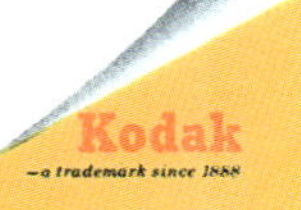

◀ 1937, USA   ▲ 1965, USA

Kodachrome Slides

Initially, Kodachrome slides were returned to the customer in the form of a single, uncut filmstrip, leaving customers responsible for cutting and mounting each image in glass for safe handling when viewing and projecting.

In 1937, Kodak introduced the affordably priced 35mm Kodaslide projector for 2" x 2" glass-mount slides and Kodaslide metal frames, which were easier to handle and lighter than glass mounts.

The big breakthrough for consumers came in April 1939 when Kodak began cutting 35mm filmstrips into individual slides and sealing them in cardboard Kodaslide Ready-Mount slide mounts before mailing the box of slides back to the customer. Photographers could now go from the mailbox to the slide projector in less than a minute.

These sturdy 2" x 2" cardboard mounts, which were designed by Dr. H.C. Staehle of the Kodak Research Laboratory, were sequentially numbered and featured a red border on the emulsion side, which was the side that faced the screen when projected.

To better protect the slide against scratches, a clear lacquer coating was also applied to the emulsion side of the film at this time. The practice of applying lacquer was discontinued in 1957, when improved gelatin hardeners provided sufficient abrasion protection.

Cutting and mounting the individual slides in Ready-Mounts or comparable cardboard and plastic alternative mounts from other manufacturers remains standard procedure to this very day among film processors worldwide.

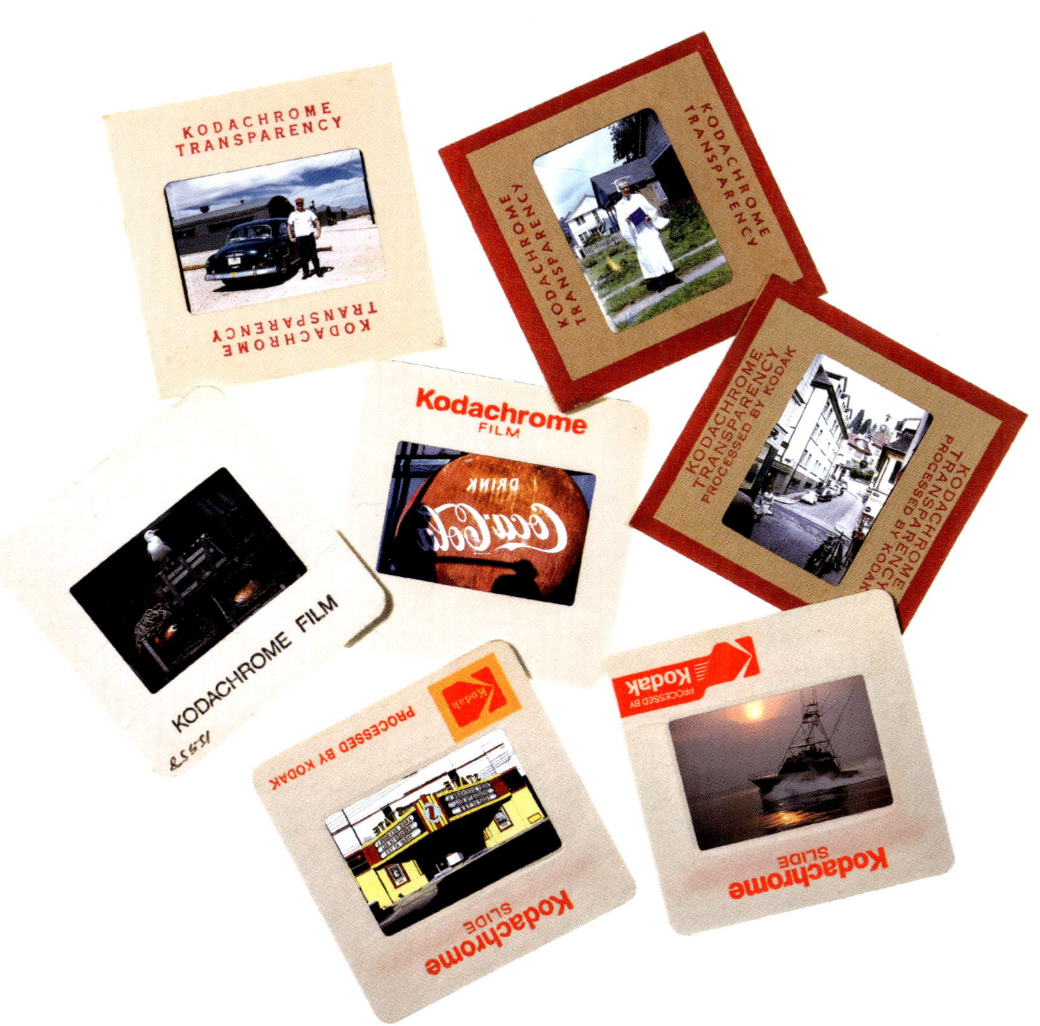

KODACHROME
TRANSPARENCY
KODACHROME
TRANSPARENCY
KODACHROME
TRANSPARENCY
KODACHROME
TRANSPARENCY
Kodachrome
FILM
KODACHROME FILM
KODACHROME
TRANSPARENCY
PROCESSED BY KODAK
KODACHROME
TRANSPARENCY
PROCESSED BY KODAK
DRINK
Coca-Cola
PROCESSED BY KODAK
Kodak
Kodachrome
SLIDE
Kodak
PROCESSED BY
Kodachrome
SLIDE

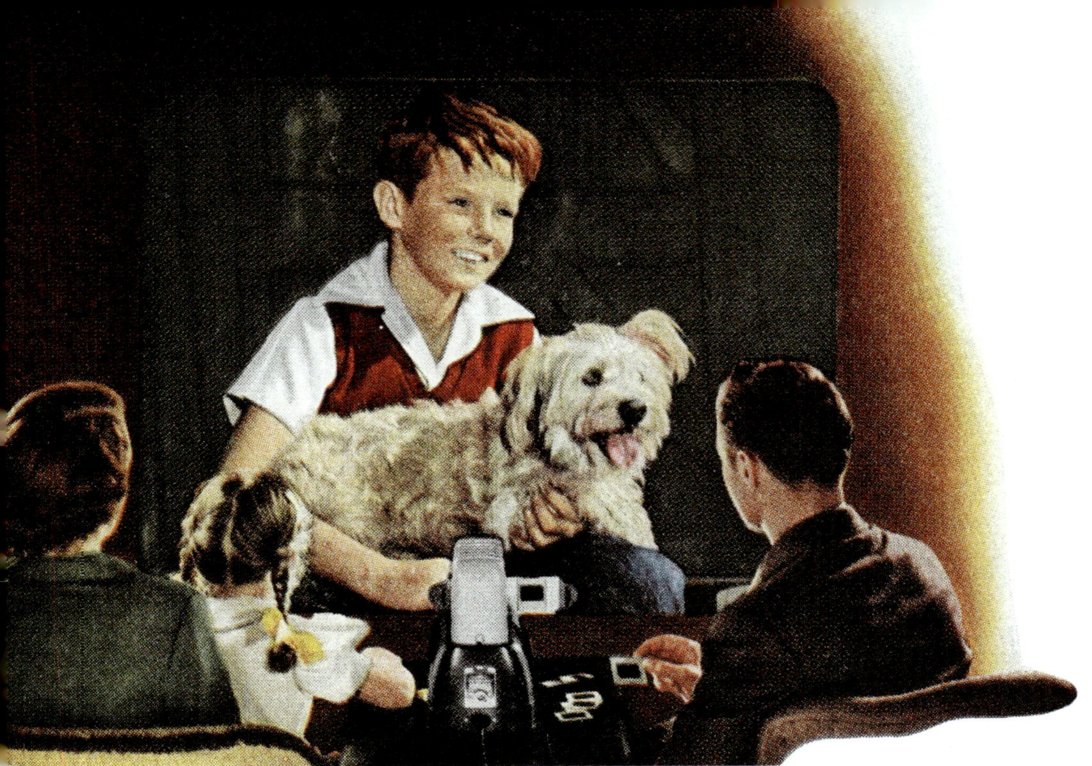

If there is a downside to 35mm slides, it would have to be that the entire viewing area is only about one square inch in size (24 x 36mm), making them difficult to view and truly appreciate. To get around the problem, Kodak and other manufacturers began producing slide viewers varying in sizes, from small pocket-size folding viewers to larger electric backlit models. In time, Kodak and others began production of Kodak Carousel Slide projectors, which became increasingly sophisticated with each new model. Projection lamps became brighter and zoom lenses were introduced for better control of image size. Slide carousels capable of holding up to eighty* or 140 slides replaced single-slot projectors, while autofocus guaranteed sharper image quality when projecting slides.

*The eighty-capacity Carousel slide trays were preferable because they were less prone to jamming. They also accepted slides mounted in sturdier (though thicker) glass and plastic slide mounts.

◀ Kodachrome ad depicting a family using
a Kodaslide projector (1947, USA)

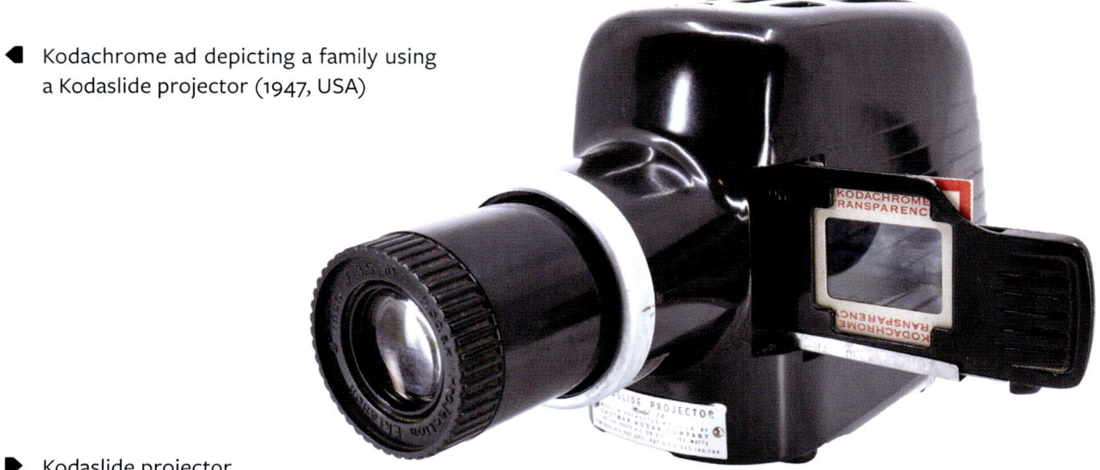

▶ Kodaslide projector

▼ Assortment of pocket slide viewers
(1930s-1940s)

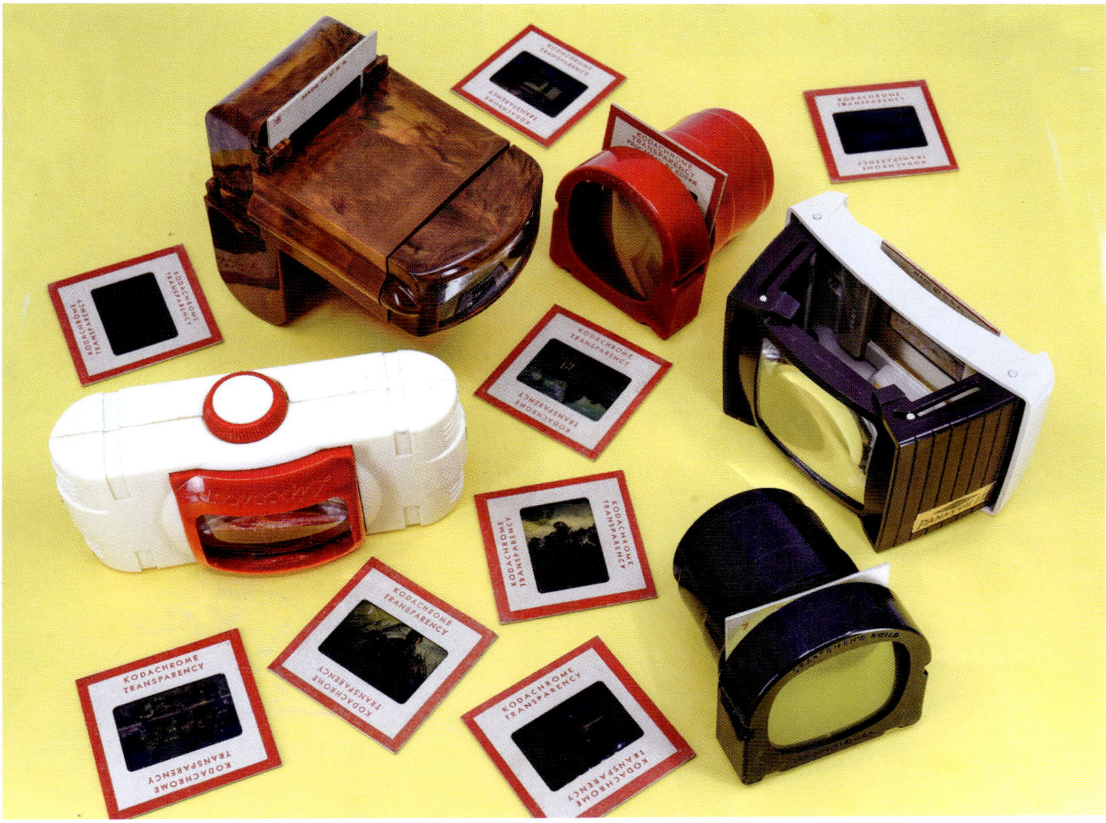

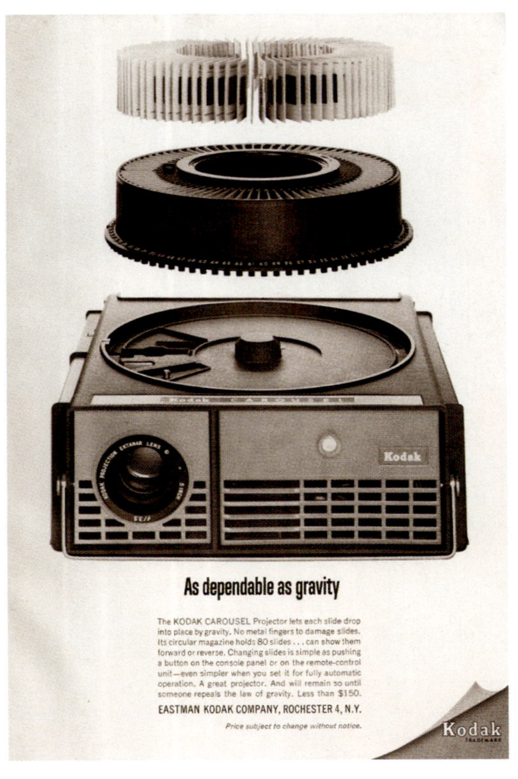

1963, USA

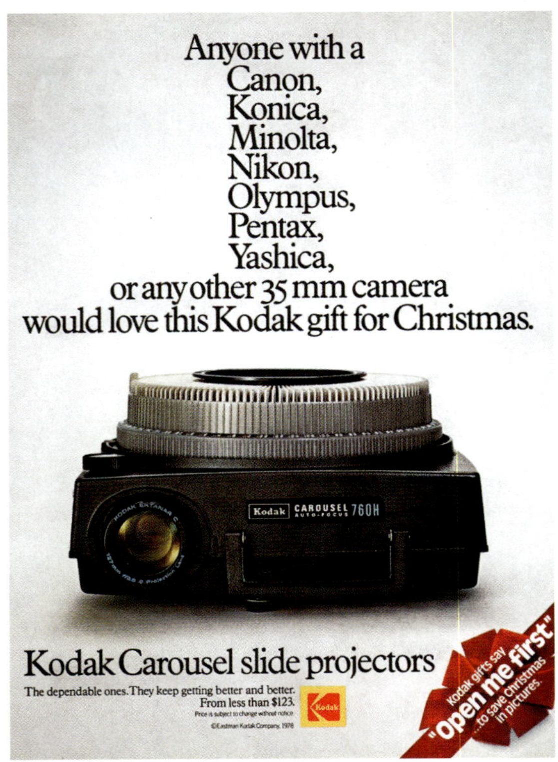

1978, USA

1983, USA

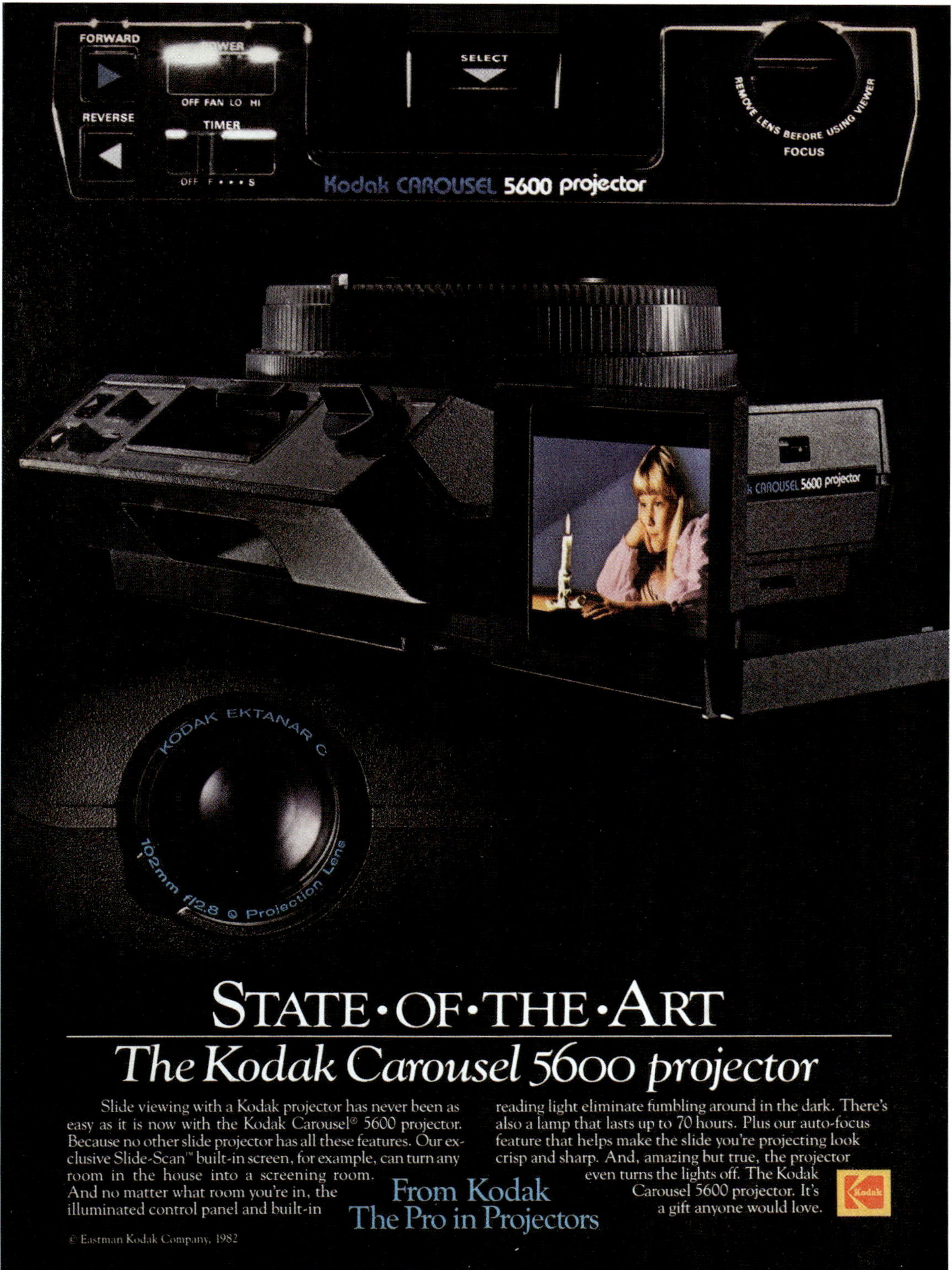

STATE·OF·THE·ART

The Kodak Carousel 5600 projector

Slide viewing with a Kodak projector has never been as easy as it is now with the Kodak Carousel® 5600 projector. Because no other slide projector has all these features. Our exclusive Slide-Scan™ built-in screen, for example, can turn any room in the house into a screening room. And no matter what room you're in, the illuminated control panel and built-in reading light eliminate fumbling around in the dark. There's also a lamp that lasts up to 70 hours. Plus our auto-focus feature that helps make the slide you're projecting look crisp and sharp. And, amazing but true, the projector even turns the lights off. The Kodak Carousel 5600 projector. It's a gift anyone would love.

**From Kodak
The Pro in Projectors**

© Eastman Kodak Company, 1982

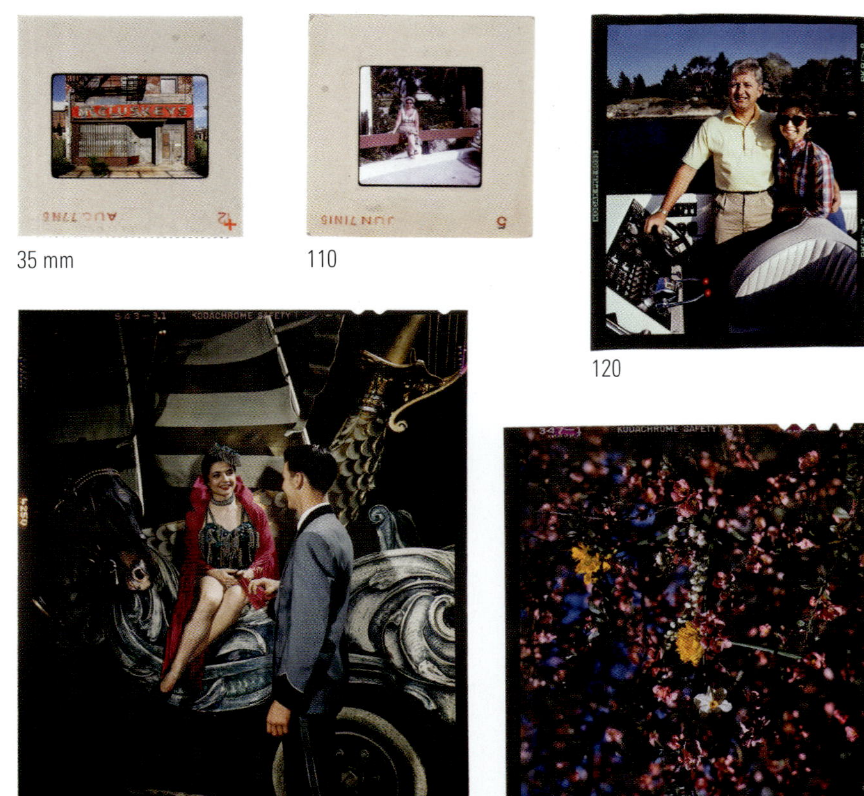

35 mm

110

120

4"x 5"

3 1/4"x 4 1/4"

The first thing that comes to most people's minds when you mention the name Kodachrome is "35mm slides." The truth is, over the course of seventy-four years, Kodachrome was available in the form of 828, 35mm, 110, 126, and 120 rolls as well as 3 ¼" x 4 ¼", 4 x 5", 8" x 10", and 11" x 14" sheet film. For motion-picture use, Kodachrome was also available in 8mm, Super 8, and 16mm.

More than 35mm Slides

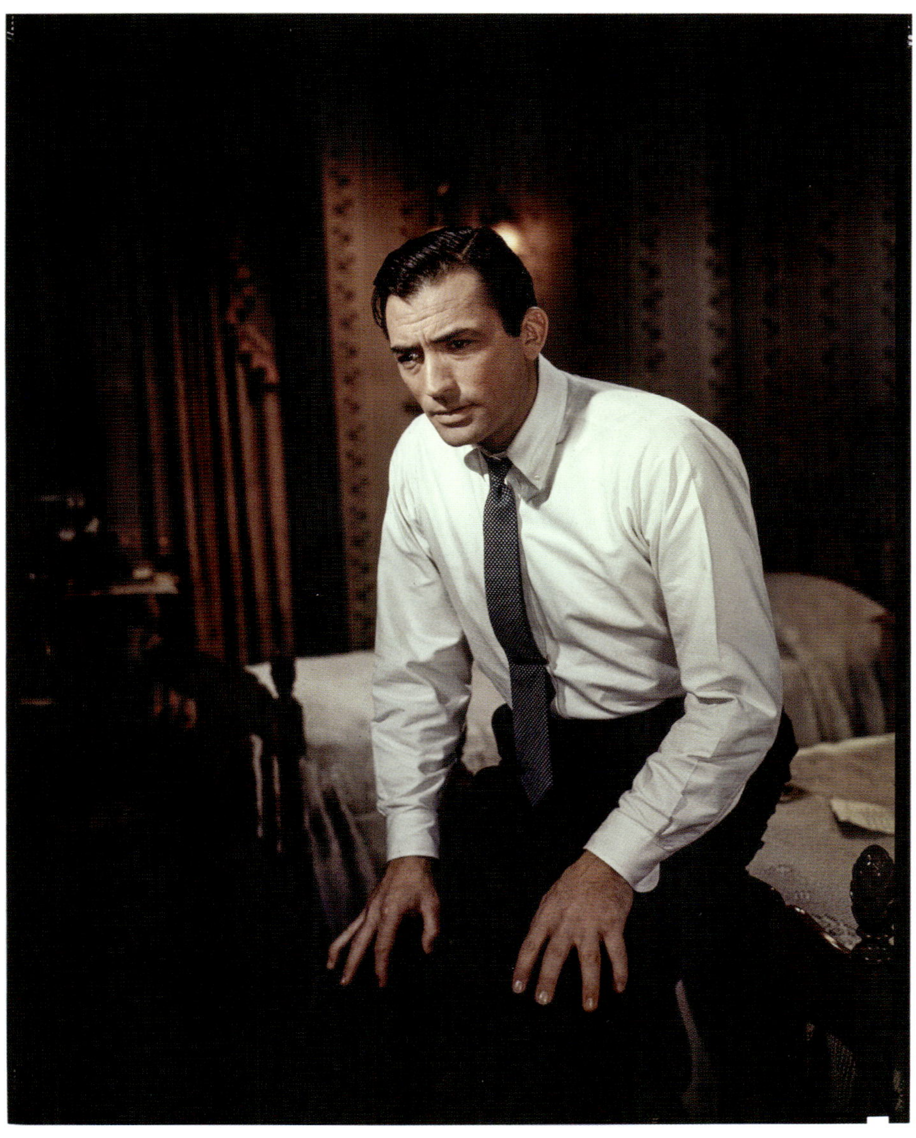

8" x 10"

Processing Mailers

The original Kodachrome might have been formulated in a bathtub but, in practice, developing Kodachrome was well beyond the capabilities of home darkrooms. As a result, Kodachrome was sold with prepaid mailers that customers would use to return the exposed film to one of Kodak's film-processing centers, which prior to the outbreak of World War II numbered forty-five in thirty-nine countries. This went on until 1954, at which time the U.S. Department of Justice threatened Kodak with antitrust violations.

As a result, Kodak agreed to allow customers to purchase Kodachrome with or without prepaid mailers. As part of the deal, Kodak also agreed to license Kodachrome development patents to independent processing labs.

By the time the sixties and seventies came around fewer than twenty Kodachrome processing labs remained operational worldwide.

In 1994, Kodak challenged the 1954 decision as being out of sync with modern business law and practice and won. As a result, Kodak regained a significant market share of the global film processing business including Kodachrome. Unfortunately for film aficionados, digital technologies were already making waves in the photography community.

Unlike the instant gratification afforded by digital cameras, Polaroid, and FUJIFILM instax print cameras, Kodachrome had to be returned to the lab for processing, which meant having to wait days or a week or more before you could kick back, settle into your favorite easy chair, and view the results.

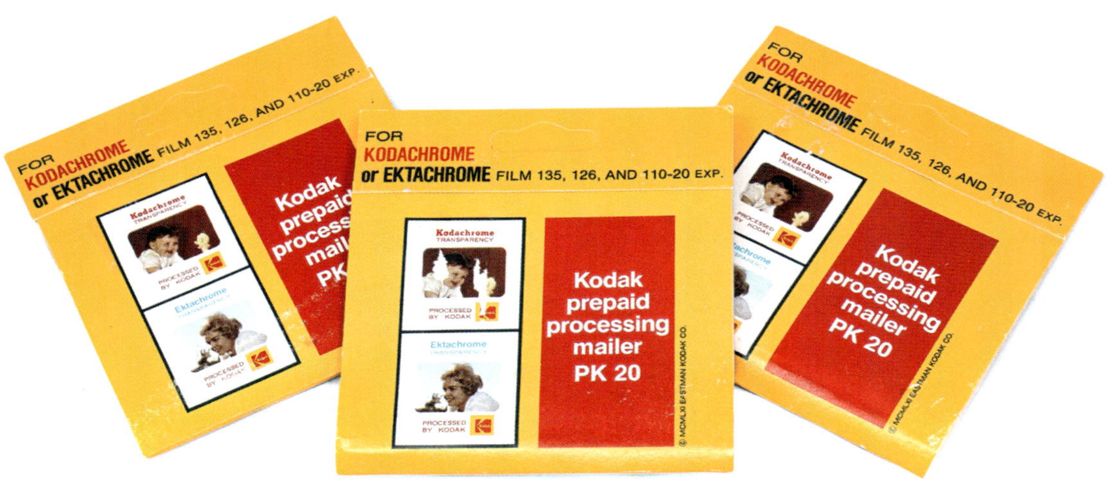

FOR
KODACHROME
or EKTACHROME FILM 135, 126, AND 110-20 EXP.
Kodachrome
TRANSPARENCY
PROCESSED BY KODAK
Ektachrome
TRANSPARENCY
PROCESSED BY KODAK
Kodak
prepaid
processing
mailer
PK 20

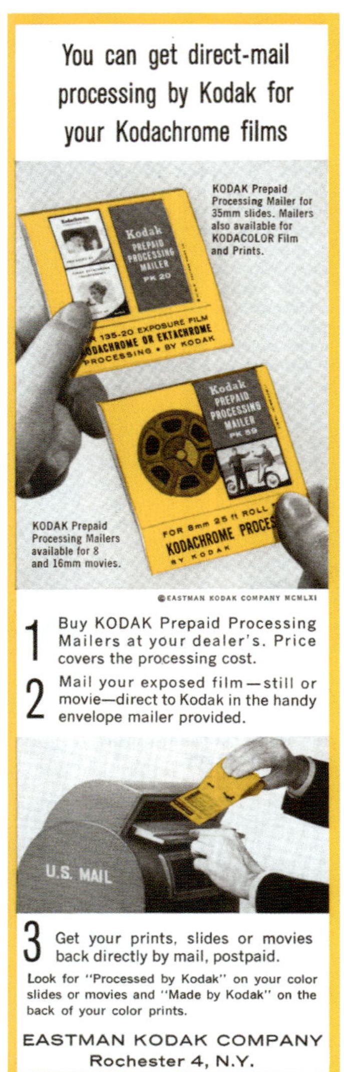

You can get direct-mail processing by Kodak for your Kodachrome films

KODAK Prepaid Processing Mailer for 35mm slides. Mailers also available for KODACOLOR Film and Prints.

Kodak PREPAID PROCESSING MAILER PK 20

FOR 135-20 EXPOSURE FILM KODACHROME OR EKTACHROME PROCESSING • BY KODAK

Kodak PREPAID PROCESSING MAILER PK 59

FOR 8mm 25 ft ROLL KODACHROME PROCES BY KODAK

KODAK Prepaid Processing Mailers available for 8 and 16mm movies.

©EASTMAN KODAK COMPANY MCMLXI

1 Buy KODAK Prepaid Processing Mailers at your dealer's. Price covers the processing cost.

2 Mail your exposed film—still or movie—direct to Kodak in the handy envelope mailer provided.

U.S. MAIL

3 Get your prints, slides or movies back directly by mail, postpaid.

Look for "Processed by Kodak" on your color slides or movies and "Made by Kodak" on the back of your color prints.

EASTMAN KODAK COMPANY
Rochester 4, N.Y.

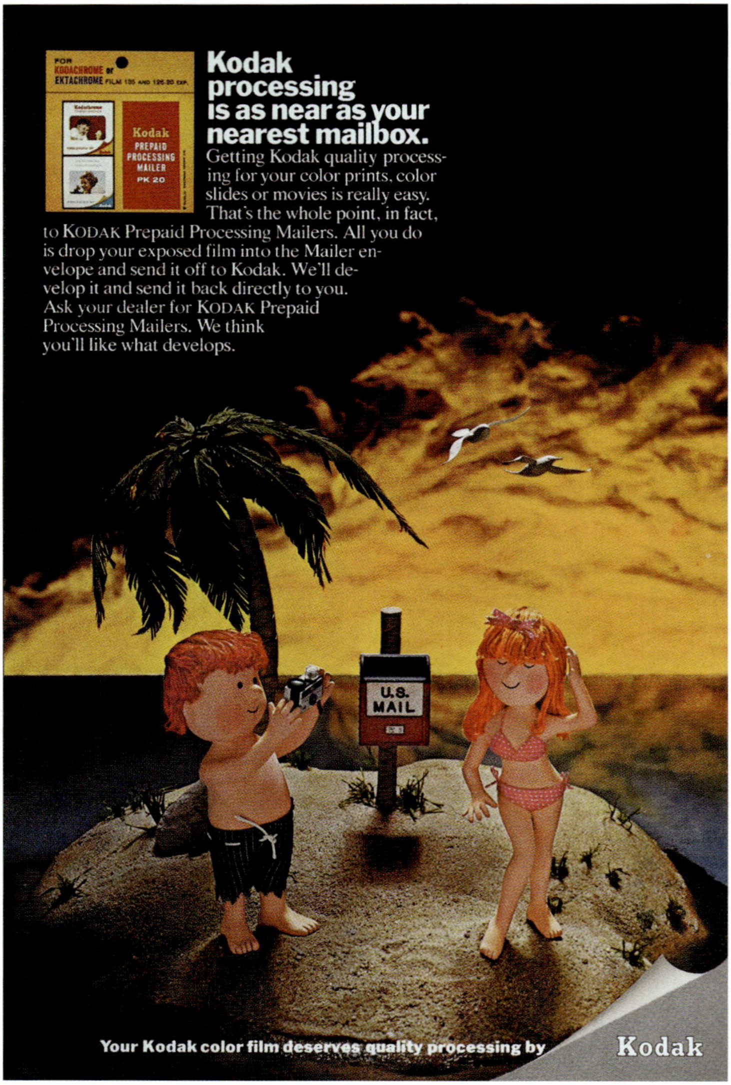

FOR KODACHROME or EKTACHROME FILM 135 AND 126-20 EXP.

Kodak PREPAID PROCESSING MAILER PK 20

Kodak processing is as near as your nearest mailbox.

Getting Kodak quality processing for your color prints, color slides or movies is really easy. That's the whole point, in fact, to KODAK Prepaid Processing Mailers. All you do is drop your exposed film into the Mailer envelope and send it off to Kodak. We'll develop it and send it back directly to you. Ask your dealer for KODAK Prepaid Processing Mailers. We think you'll like what develops.

U.S. MAIL

Your Kodak color film deserves quality processing by   Kodak

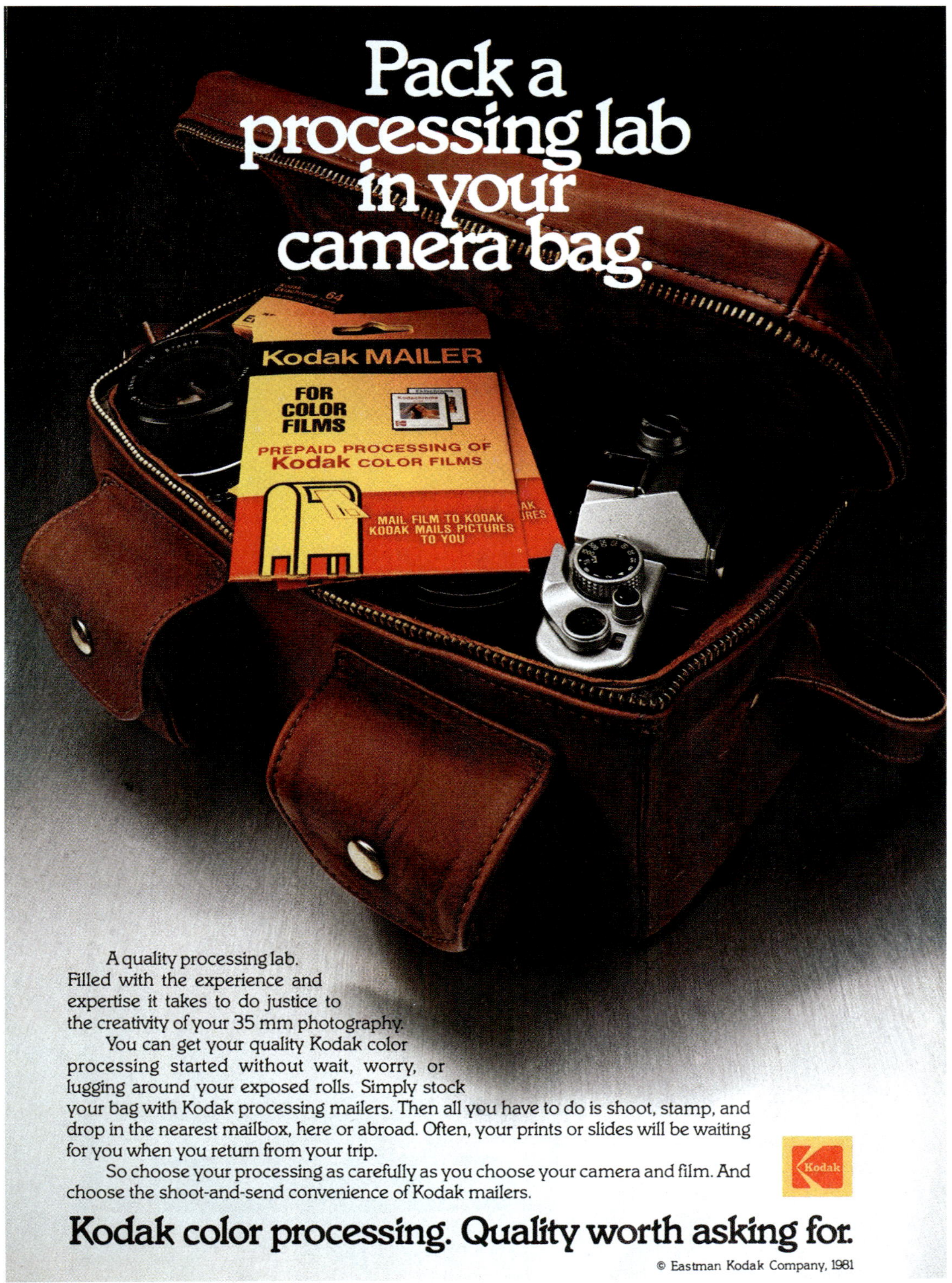

Pack a
processing lab
in your
camera bag.

Kodak MAILER
FOR COLOR FILMS
PREPAID PROCESSING OF Kodak COLOR FILMS
MAIL FILM TO KODAK
KODAK MAILS PICTURES TO YOU

A quality processing lab.
Filled with the experience and
expertise it takes to do justice to
the creativity of your 35 mm photography.
You can get your quality Kodak color
processing started without wait, worry, or
lugging around your exposed rolls. Simply stock
your bag with Kodak processing mailers. Then all you have to do is shoot, stamp, and
drop in the nearest mailbox, here or abroad. Often, your prints or slides will be waiting
for you when you return from your trip.
So choose your processing as carefully as you choose your camera and film. And
choose the shoot-and-send convenience of Kodak mailers.

Kodak color processing. Quality worth asking for.

Kodak

© Eastman Kodak Company, 1981

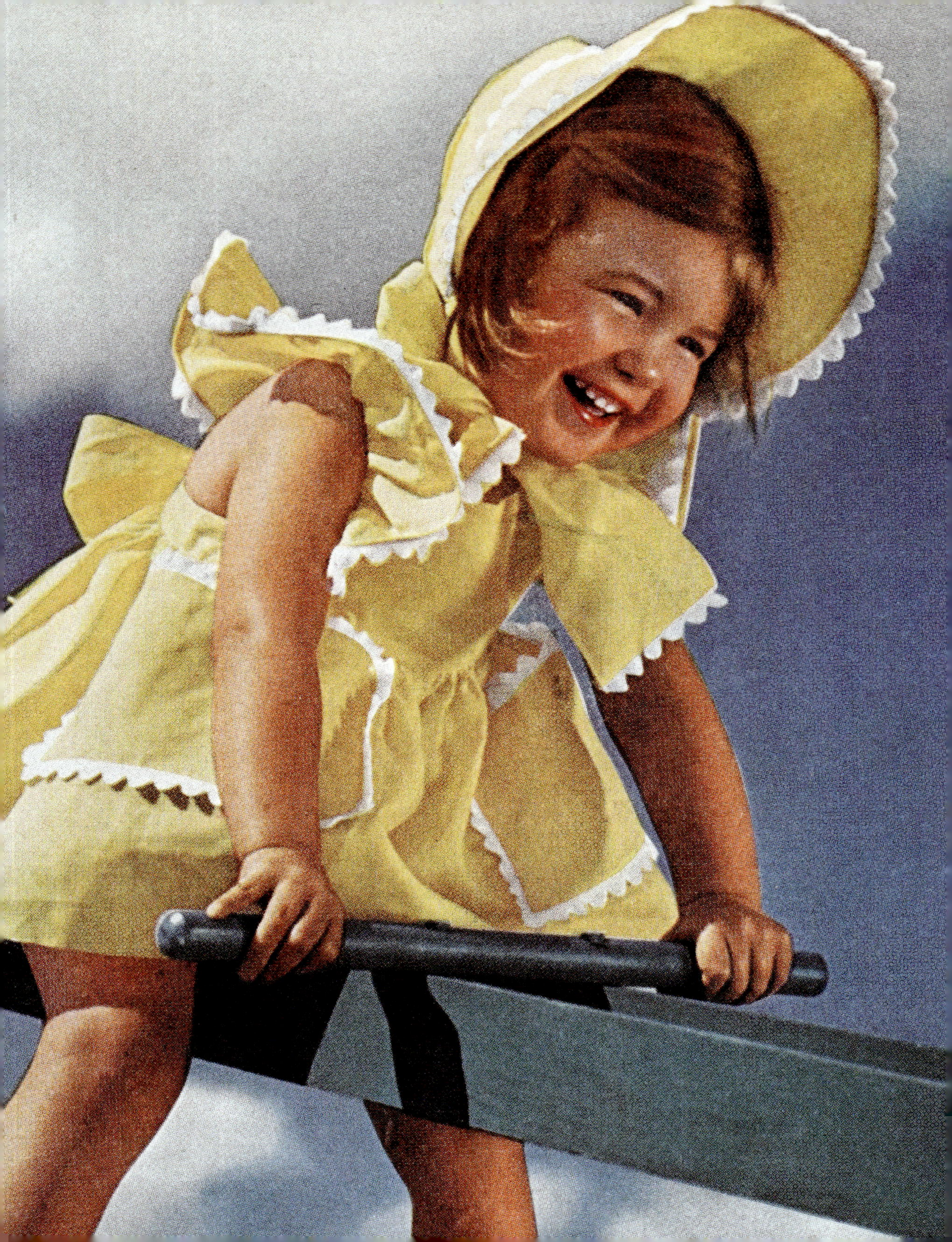

SELLING COLOR

Kodachrome Ads, 1935-1991

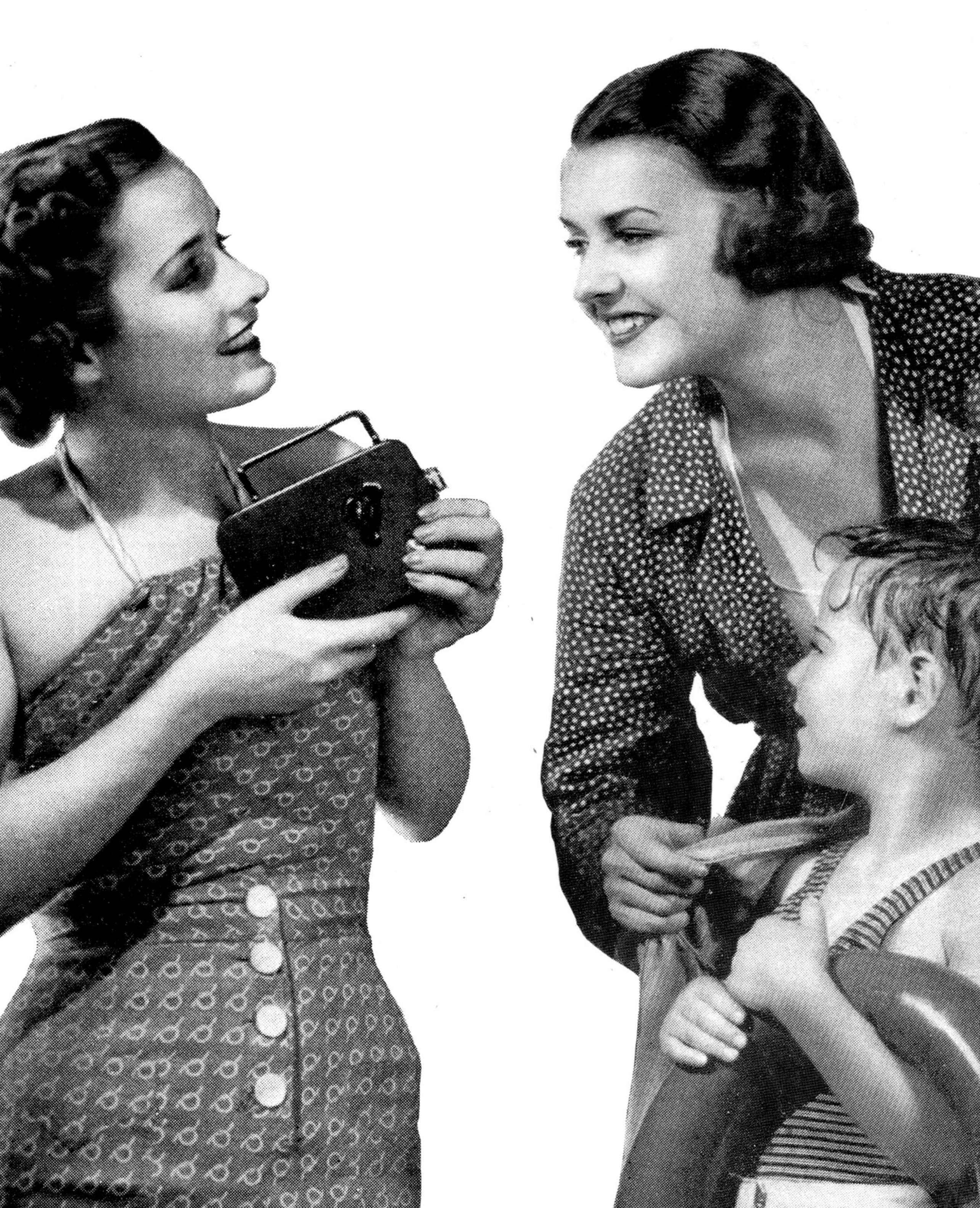

Introducing Kodachrome
1935-1941

Within the first year of Kodachrome's introduction, Kodak began aggressively promoting the new film, along with all of the cameras, projectors, and accessories made to pair with it. J Walter Thompson, Kodak's advertising agency from 1930 to 1997, was responsible for Kodak's focused sense of mission when it came to promoting the company's "Every moment counts" marketing theme. Curiously, many of the earliest US Kodachrome print advertisements were printed in black-and-white—not in color. This would soon change.

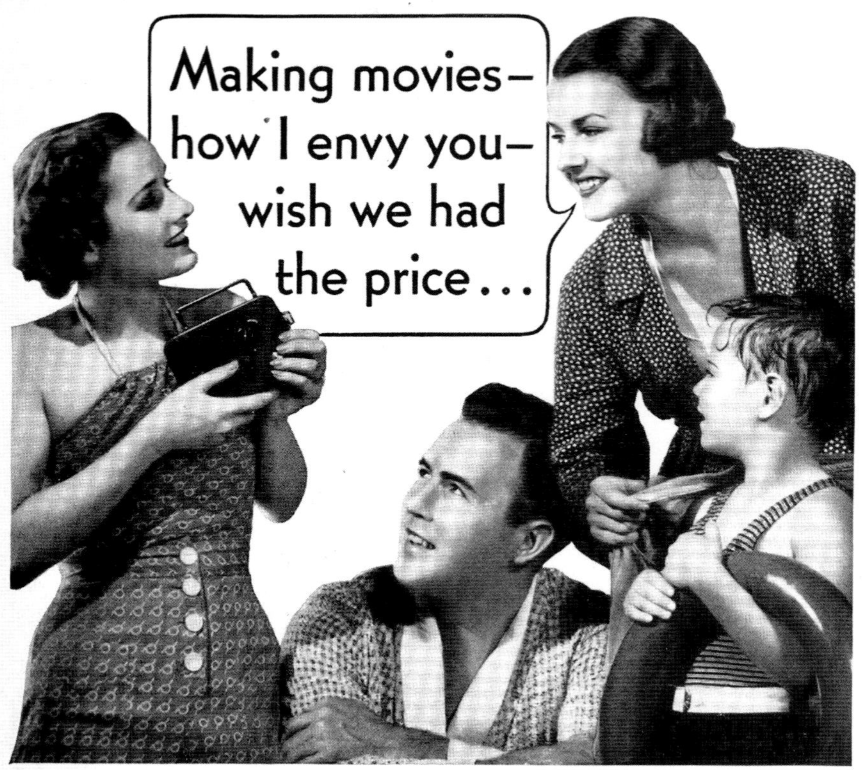

. . . You have—CINÉ-KODAK EIGHT was made for movie makers who haven't a lot of money

A FASCINATING SPORT, all right, this making home movies, but a bit on the expensive side. Is that what you've been thinking? Then it's high time you knew about Ciné-Kodak Eight—a full-fledged camera — the home movie maker designed for people who have to watch their pennies.

A 25-foot roll of black-and-white film for Ciné-Kodak Eight costs only $2.25, *finished, ready to show*. Yet it runs as long on the screen as 100 feet of amateur standard home movie film—gives you 20 to 30 movie scenes, each as long as the average scene in the newsreels.

FULL COLOR WITH KODACHROME

Load the Eight with the remarkable new Kodachrome Film, and you can make movies in color— gorgeous full color. Simple to make as black-and-white, and the cost is just a few cents more a scene. No extra equipment needed. The color is in the film. See movies in black-and-white and in full-color Kodachrome at your dealer's . . . Eastman Kodak Company, Rochester, N. Y.

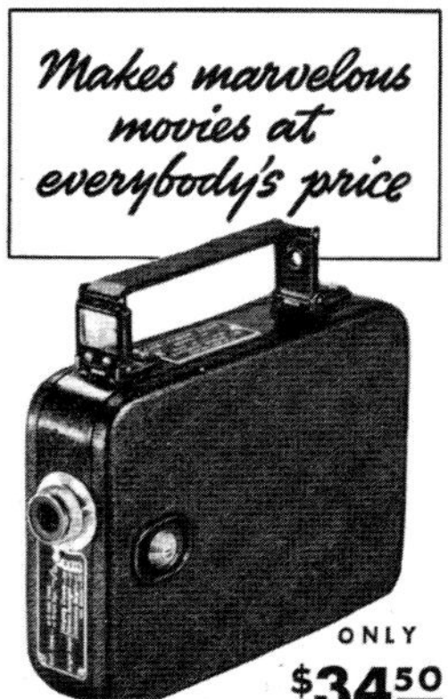

One of the earliest print advertisements for Kodachrome film was this 1935 trade advertising card (2″ x 2 ¾″), distributed by Jacques Superchocolat, a well-known Belgian chocolate maker. At the turn of the twentieth century, new consumer products were often introduced to the public through the use of trade advertising cards, which typically had a description of the product on one side and an illustration on the reverse side—though not necessarily of the advertised product.

For reasons unknown, a watercolor rendering of a church in an Alpine village appears on the picture side of this card rather than a colorful photograph taken with Kodachrome. On the reverse side are the logo and address of Jacques Superchocolat along with a brief description of Kodak's new color film, in French and Dutch.

▶ 1935, Belgium

◀ 1936, USA

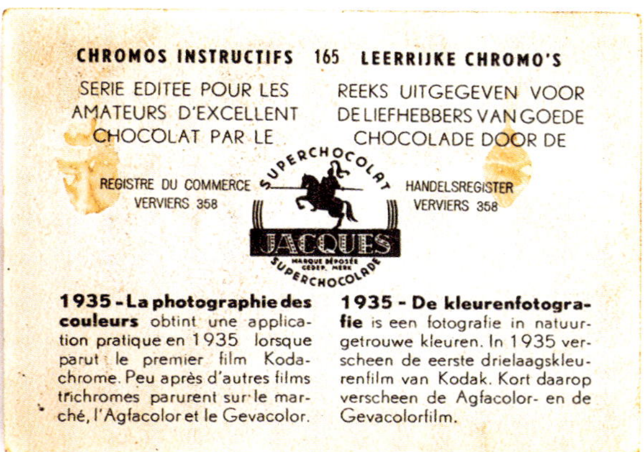

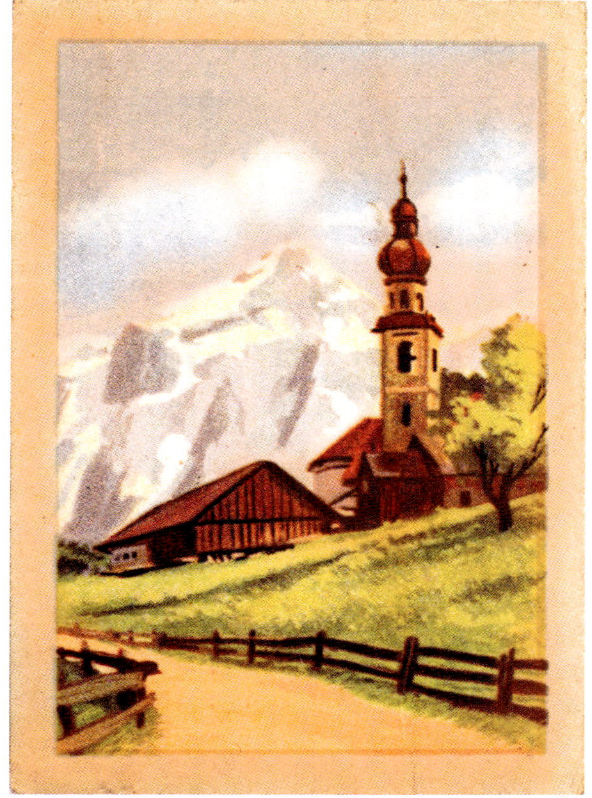

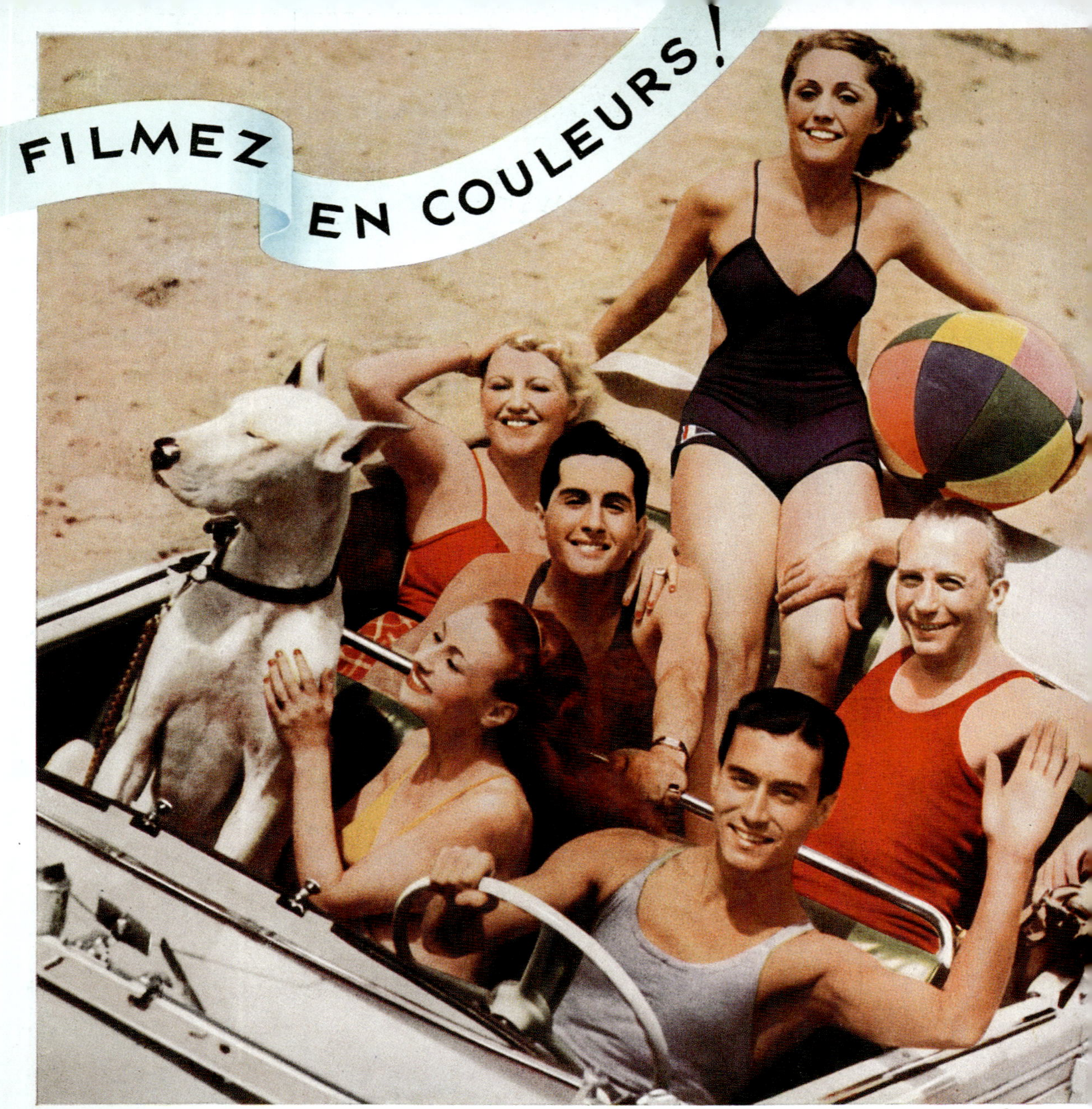

De telles scènes colorées, vous en tournerez à votre guise avec le nouveau film "Kodachrome" qui se place sur n'importe quel appareil de prises de vues, en 16 et 8 m/m.

Désormais, avec l'action, c'est le bleu du ciel, la carnation des visages, l'or des chevelures, la bigarrure des robes, des fleurs, que vous projetterez sur votre écran.

Et tourner en couleurs est aussi facile que de tourner en noir. Pas d'accessoires supplémentaires. Le film "Kodachrome" fait tout.

Goûtez aux joies du cinéma en couleurs naturelles.

Offrez à vos amis ce régal des yeux.

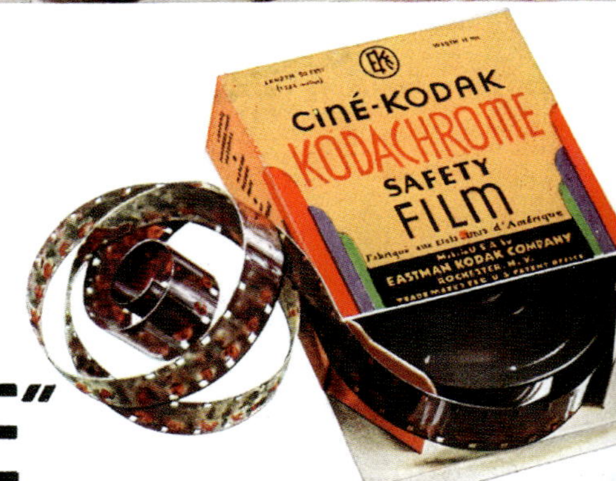

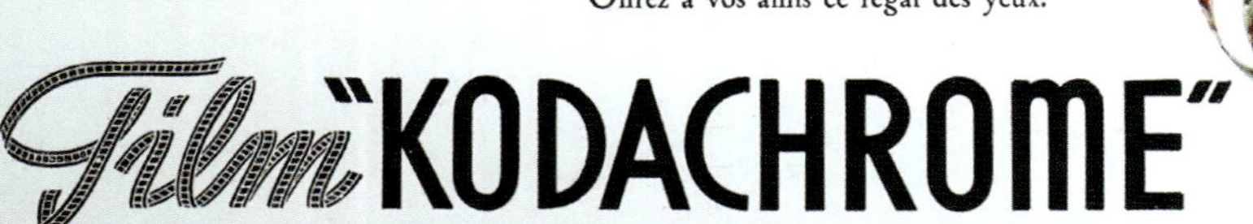

While Kodak was publishing staid and monochromatic print advertisements, European agencies ditched the business suits and sensible dresses, loaded up their cameras with Kodachrome, and headed for the beach. This was especially true for advertising and magazines targeting the French-speaking market. European agencies wisely chose to emphasize the color qualities of Kodachrome without ignoring the ease-of-use aspects of Kodak cameras and accessories. The resulting ads remain visually fresh and fun.

◀ 1936, France

▶ 1936, France

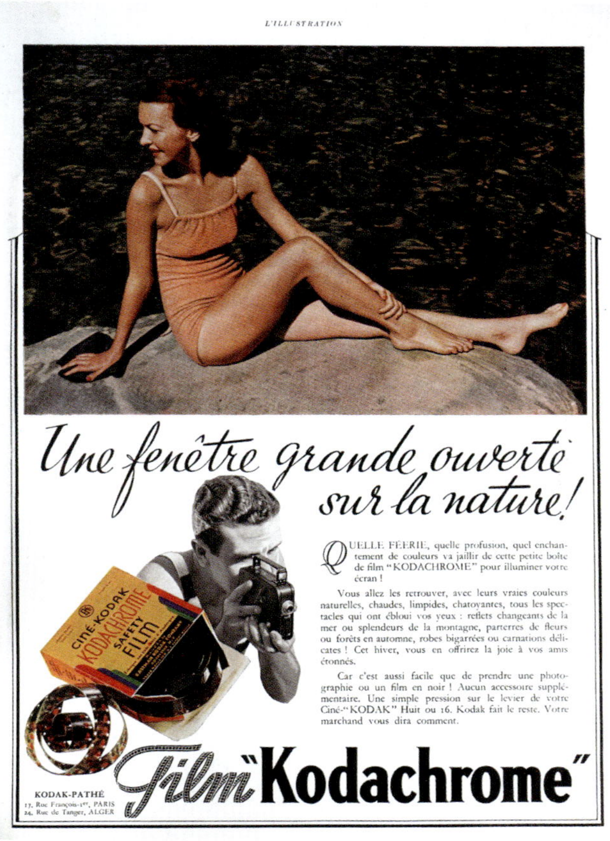

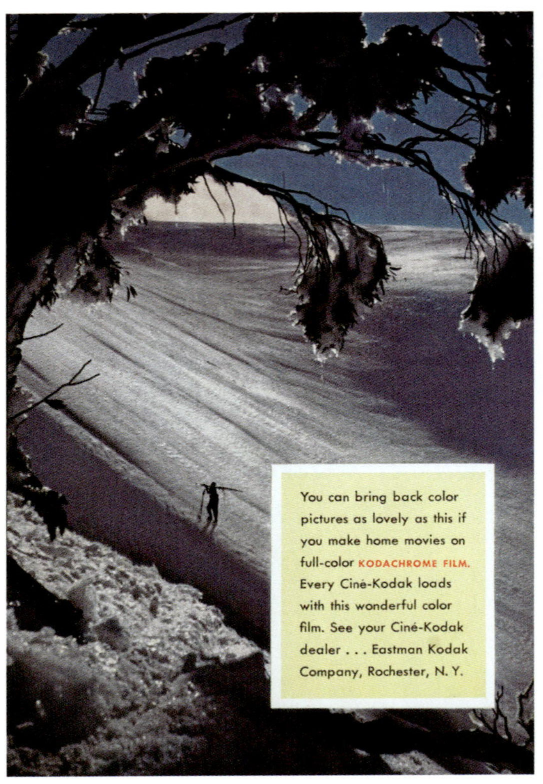

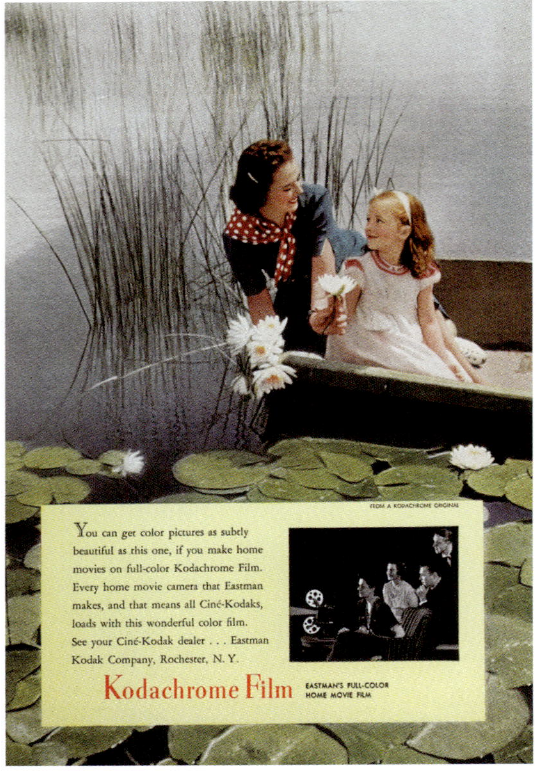

From 1940 through 1942, Kodak's US print advertisements start hitting their stride. Gone were the monochromatic, parochial look of earlier ads. Kodachrome is about color and life, and that's exactly what these advertisements illustrated.

1940, USA ▲

1940, USA ▶

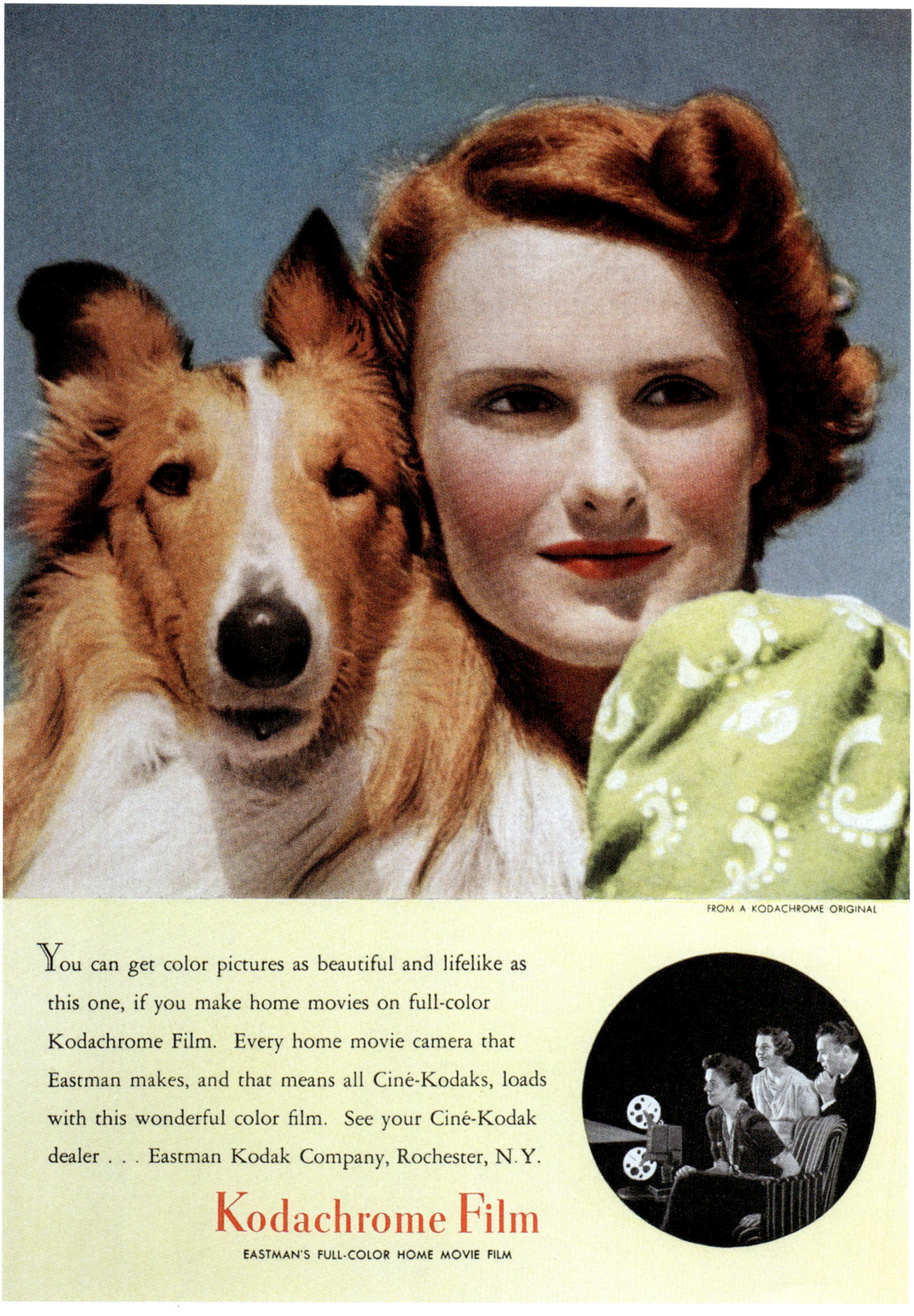

You can get color pictures as beautiful and lifelike as this one, if you make home movies on full-color Kodachrome Film. Every home movie camera that Eastman makes, and that means all Ciné-Kodaks, loads with this wonderful color film. See your Ciné-Kodak dealer . . . Eastman Kodak Company, Rochester, N.Y.

Kodachrome Film

EASTMAN'S FULL-COLOR HOME MOVIE FILM

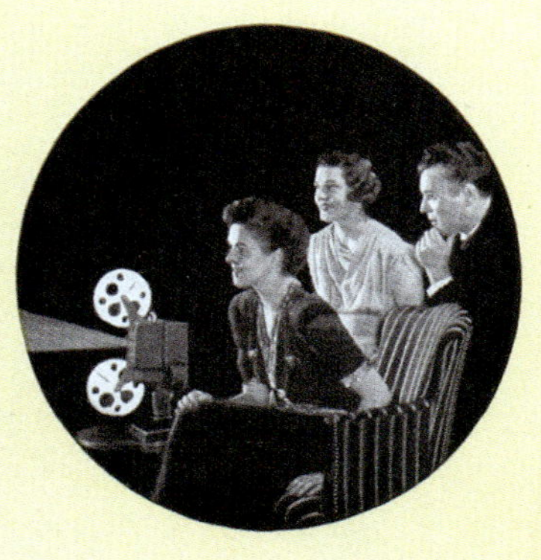

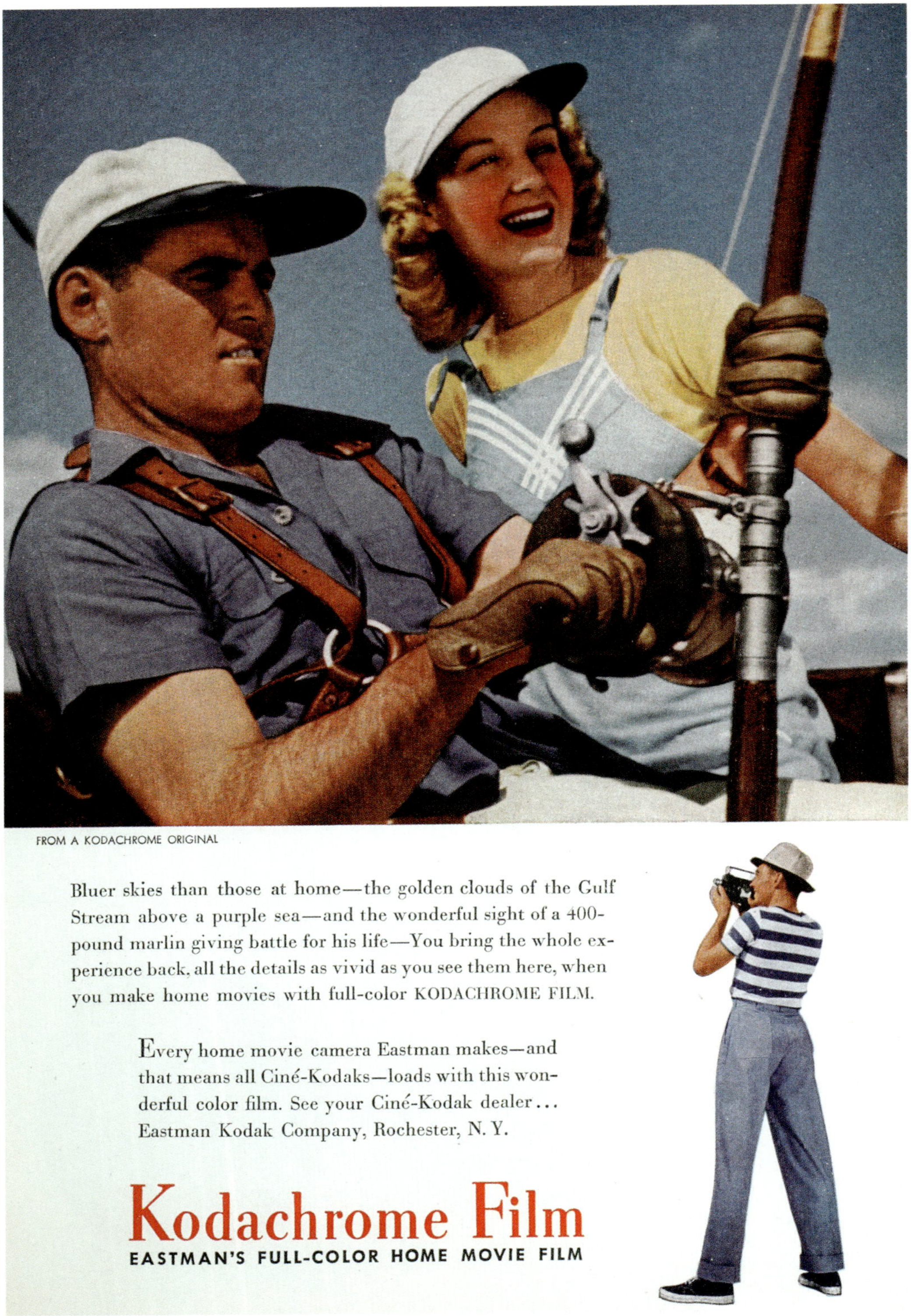

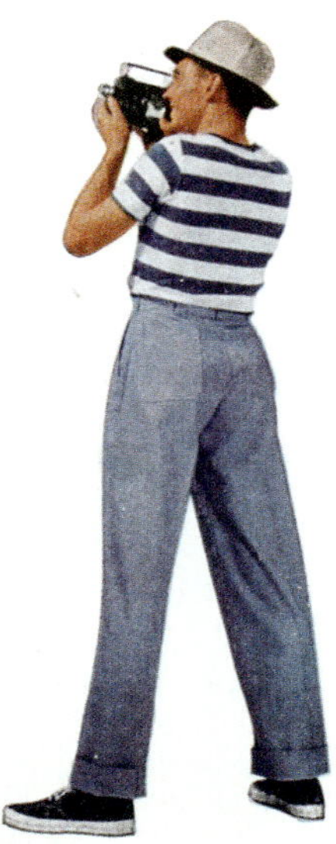

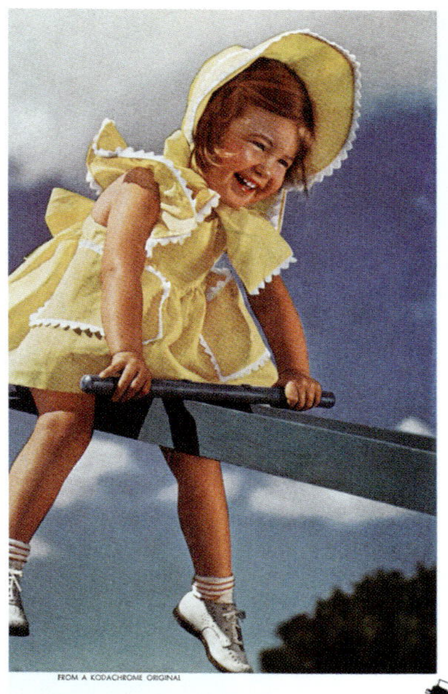

Like a wind-blown flower against the deep blue sky— flying brown hair, flying bright pinafore, round, rosy cheeks, laughing eyes—you get her, just as you see her here, when you make home movies with full-color KODACHROME FILM.

Every home movie camera Eastman makes—and that means all Ciné-Kodaks—loads with this wonderful color film. See your Ciné-Kodak dealer . . . Eastman Kodak Company, Rochester, N. Y.

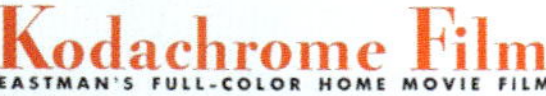

Kodachrome Film
EASTMAN'S FULL-COLOR HOME MOVIE FILM

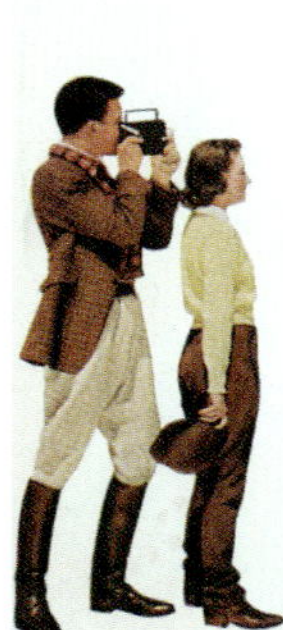

A sky as brilliant as a jewel—the burning gold of autumn foliage—and a couple of horseback riders, clear-cut as if cast in bronze, silhouetted against the sweep of country. You get the whole scene, with all its intense color and life, just as you see it here, when you make home movies on full-color KODACHROME FILM.

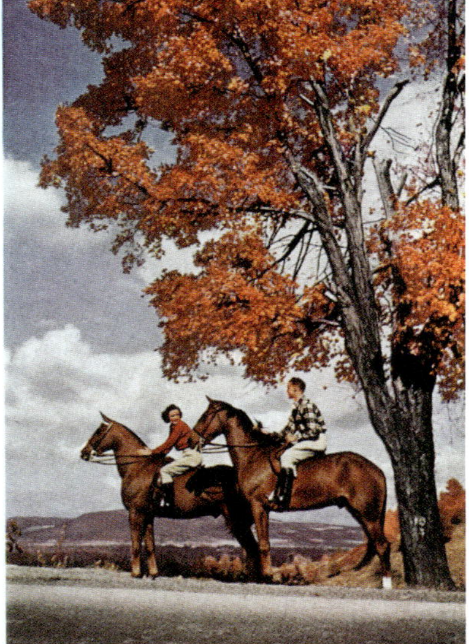

Every home movie camera Eastman makes—and that means all Ciné-Kodaks—loads with this wonderful color film. See your Ciné-Kodak dealer . . . Eastman Kodak Company, Rochester, N. Y.

Kodachrome Film
EASTMAN'S FULL-COLOR HOME MOVIE FILM

◀ 1942, USA

▲ 1941, USA

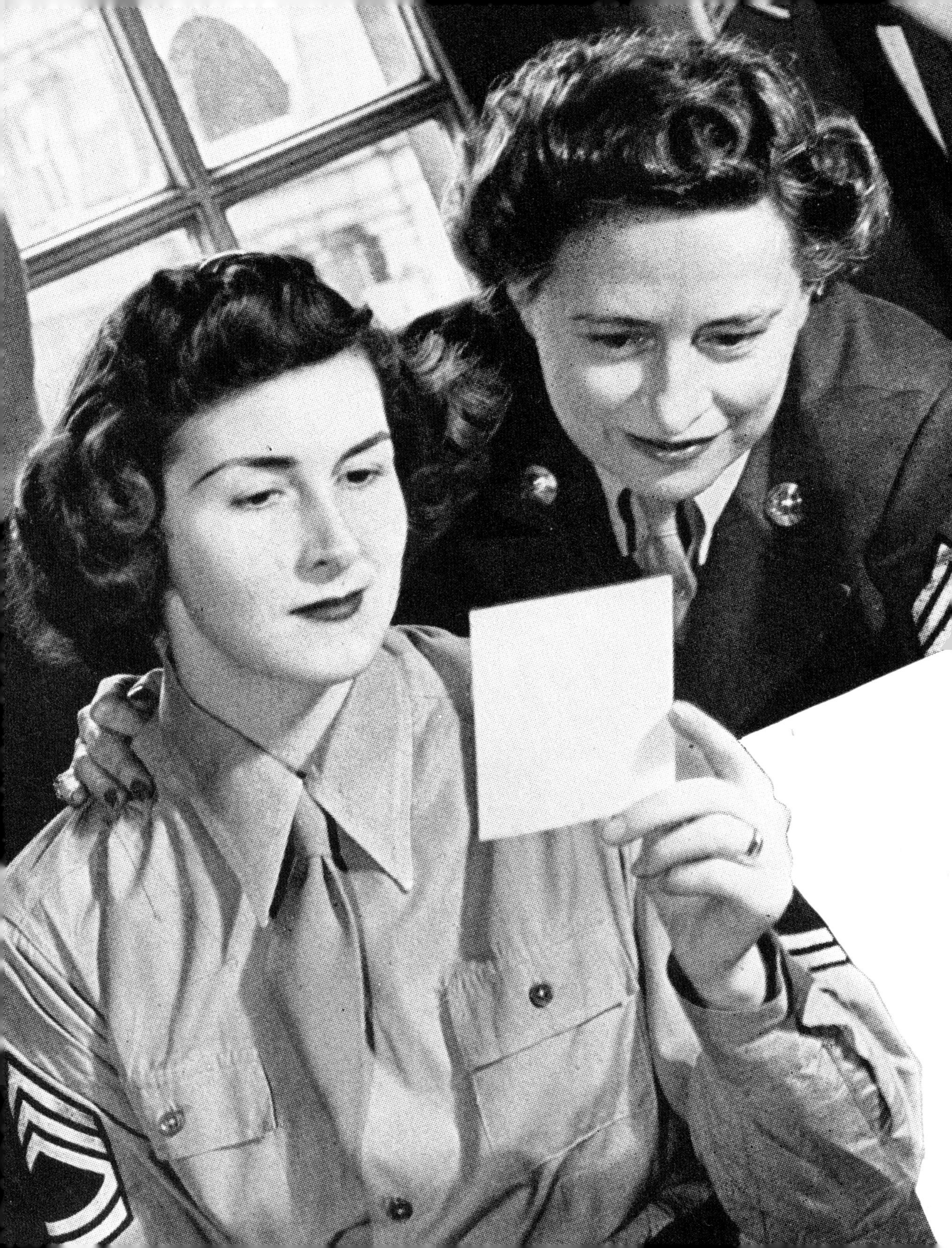

The War Years

1942-1945

The clamour for Kodachrome quickly dimmed when the United States entered World War II. With the nation at war, Kodak shifted priorities by cutting back on consumer products, including Kodachrome, in order to increase production of X-Ray film, chemistry, and other war-related provisions.

With tens of thousands of soldiers off fighting wars in distant lands, home movies were not high on the list of consumer priorities. For now, memories would have to wait. In a bid to ease the stress of the times, Kodak always made sure to include a positive, reassuring note within the copy of its wartime advertising. The promise? Good times will return "when film comes back…"

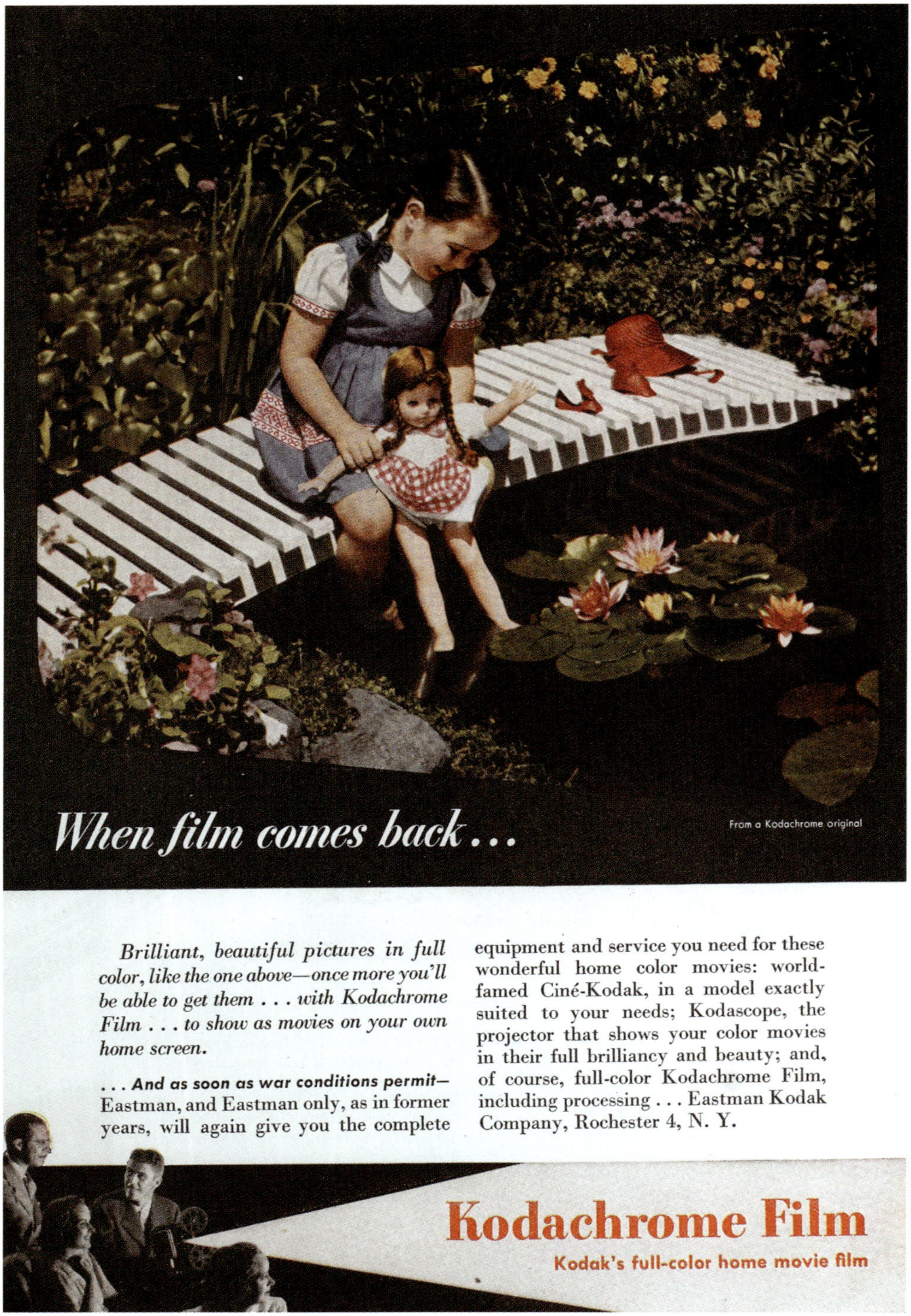

When film comes back . . .

From a Kodachrome original

Brilliant, beautiful pictures in full color, like the one above—once more you'll be able to get them . . . with Kodachrome Film . . . to show as movies on your own home screen.

. . . And as soon as war conditions permit— Eastman, and Eastman only, as in former years, will again give you the complete equipment and service you need for these wonderful home color movies: world-famed Ciné-Kodak, in a model exactly suited to your needs; Kodascope, the projector that shows your color movies in their full brilliancy and beauty; and, of course, full-color Kodachrome Film, including processing . . . Eastman Kodak Company, Rochester 4, N. Y.

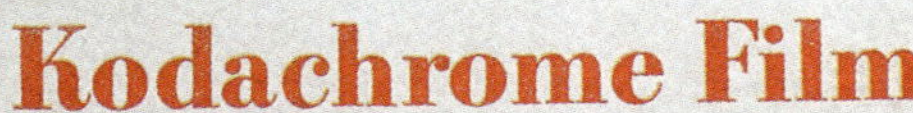

1945, USA   1944, USA

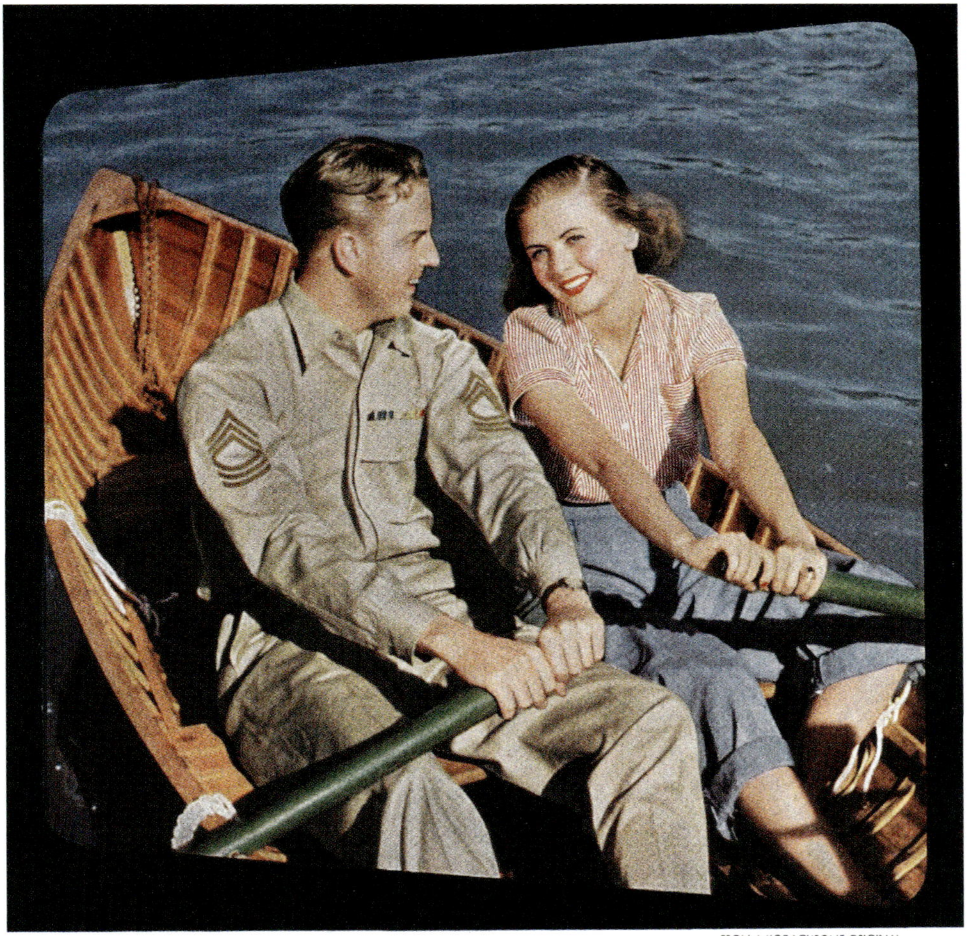

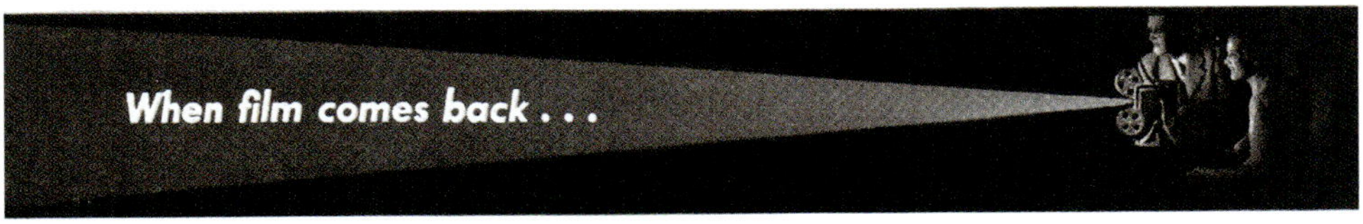

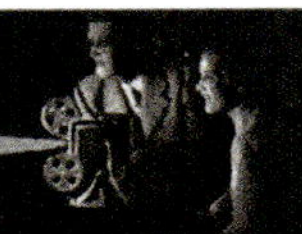

You can get color pictures as beautiful and precious as this one . . . with Kodachrome Film . . . to show as movies on your own home screen.

. . . And as soon as war conditions permit— Eastman, and Eastman only, as in former years, will again give you the complete equipment and service you need for these wonderful home color movies: world-famed Ciné-Kodak, in a model exactly suited to your needs; Kodascope, the projector that shows your color movies in their full brilliancy and beauty; and, of course, full-color Kodachrome Film, including processing . . . Eastman all, and all designed to work together . . . Eastman Kodak Co., Rochester, N. Y.

Kodachrome Film
Kodak's full-color home movie film

Photo of Saturn

The egos of geniuses can get prickly at times. If you read the text of the 1946 print ad on the opposite page, you will quickly note it contains a backhanded compliment from one Kodak division to another.

The advertisement features an impressive photograph of the planet Saturn and its rings, captured on Kodachrome through the telescope at the Mount Wilson Observatory in California.

While the text starts out praising the "weird and beautiful" photograph and what a "great achievement" the photograph is, by the start of the second paragraph the text quickly pivots to a "brains or beauty" manifesto, touting how much more impressive and scientifically valuable the black-and-white spectrographic atmospheric analysis strip, superimposed just below the photograph of the planet's rings, is for scientists.

What the guys in the lab coats are basically saying to the folks in the marketing division is that without the inclusion of the spectrograph, it's all just an advertisement with a "pretty picture of Saturn"… and have a nice day.

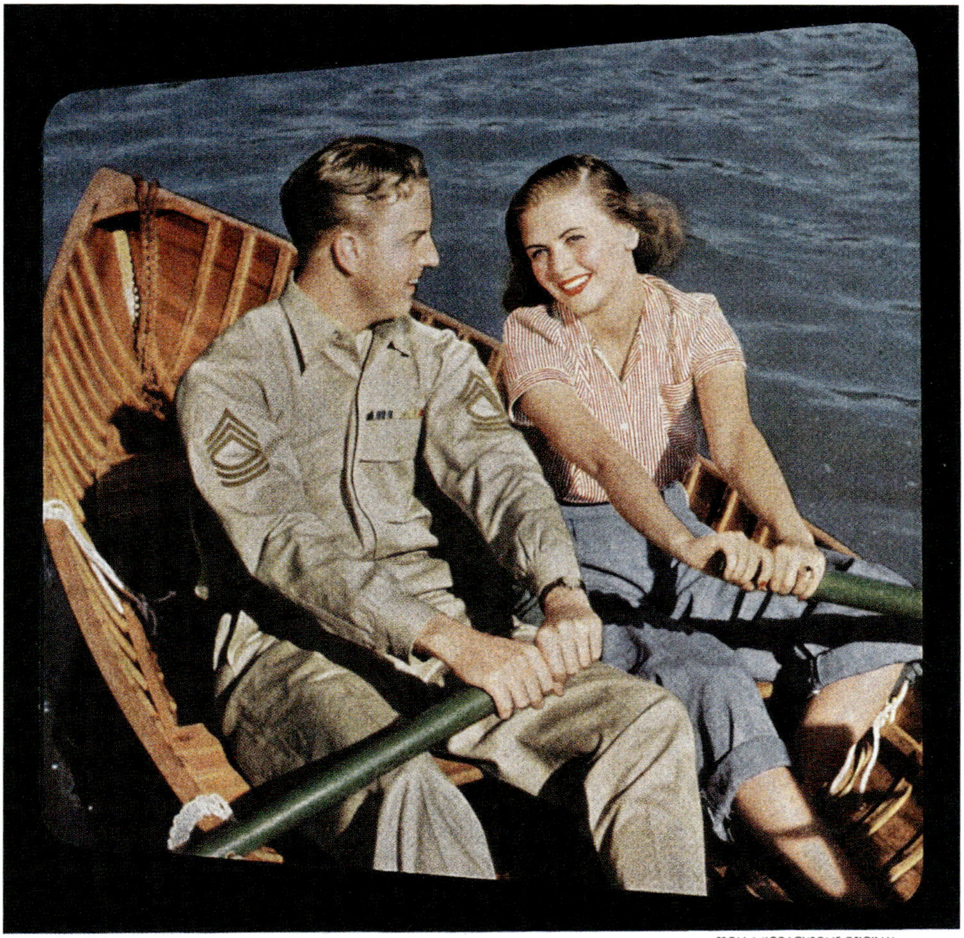

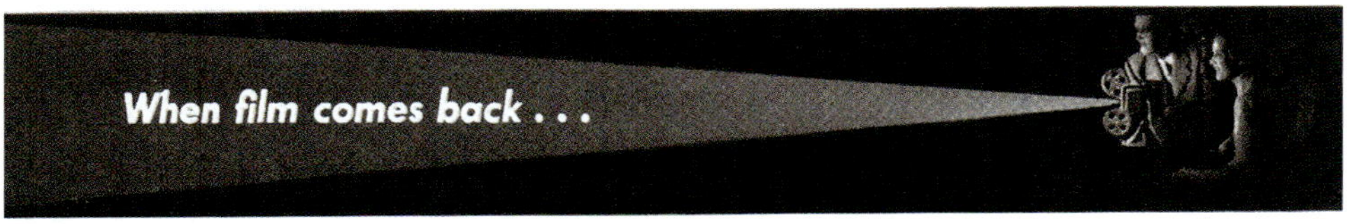

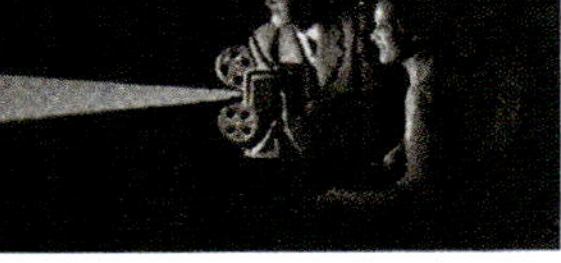

You can get color pictures as beautiful and precious as this one . . . with *Kodachrome Film* . . . to show as movies on your own home screen.

. . . And as soon as war conditions permit— Eastman, and Eastman only, as in former years, will again give you the complete equipment and service you need for these wonderful home color movies: world-famed Ciné-Kodak, in a model exactly suited to your needs; Kodascope, the projector that shows your color movies in their full brilliancy and beauty; and, of course, full-color Kodachrome Film, including processing . . . Eastman all, and all designed to work together . . . Eastman Kodak Co., Rochester, N. Y.

Kodachrome Film
Kodak's full-color home movie film

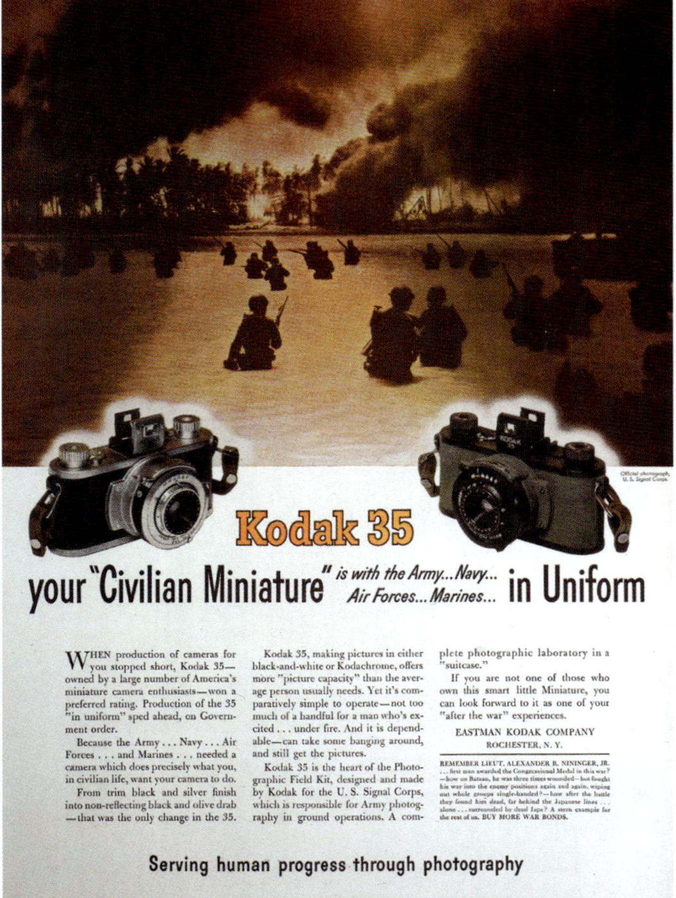

In a bid to gather support for our troops fighting overseas early on in the war, the US military hired five of the top producers and directors in Hollywood at the time, including John Ford, William Wyler, John Huston, Frank Capra, and George Stevens, to produce documentaries that would be shown in theaters back home.

One of these films, "The Battle of Midway," won an Academy Award in 1942 in the Short Documentary category. Using Kodachrome film and 16mm movie cameras, director Joh Ford captured actual footage of the battle that was used in the documentary.

1944, USA ◀

1944, USA ▶

Man in the re-making

X-ray film has the greatest assignment
since its introduction by Kodak in 1914

In its way, this picture represents a sort of miracle—symbol-
izes the tens of thousands of times in which X-rays have served
as "blueprints"... for the re-making of men.

Evidence is seen in military hospitals and in the wounded
men returned to daily life. Thousands have already been re-
stored to useful activity—a matchless tribute to this war's doctors and nurses . . . to the drugs and implements they use.

Radiography—photography on X-ray film—is the implement
with which doctors survey hidden damage, plan a course of
action, and follow the healing which surgery began.

It is the difference between finding your way in the dark,
and seeing. X-ray film has reached a new climax in its life . . .
which began with its introduction by Kodak in 1914.

EASTMAN KODAK COMPANY, ROCHESTER 4, N. Y.

REMEMBER "THE DEATH
MARCH FROM BATAAN"? . . .
How after their surrender our
boys, crazed by thirst, were forced
to drink from stagnant wallows?...
How some who collapsed were
abandoned? . . . How, at the end,
5,200 Americans from Bataan and
Corregidor found death in Japa-
nese prison camps? A stern example to us at home. BUY—AND
HOLD—MORE WAR BONDS.

Serving human progress through photography

Photo of Saturn

The egos of geniuses can get prickly at times. If you read the text of the 1946 print ad on the opposite page, you will quickly note it contains a backhanded compliment from one Kodak division to another.

The advertisement features an impressive photograph of the planet Saturn and its rings, captured on Kodachrome through the telescope at the Mount Wilson Observatory in California.

While the text starts out praising the "weird and beautiful" photograph and what a "great achievement" the photograph is, by the start of the second paragraph the text quickly pivots to a "brains or beauty" manifesto, touting how much more impressive and scientifically valuable the black-and-white spectrographic atmospheric analysis strip, superimposed just below the photograph of the planet's rings, is for scientists.

What the guys in the lab coats are basically saying to the folks in the marketing division is that without the inclusion of the spectrograph, it's all just an advertisement with a "pretty picture of Saturn"… and have a nice day.

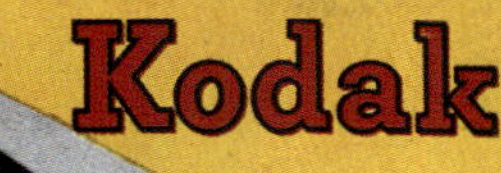

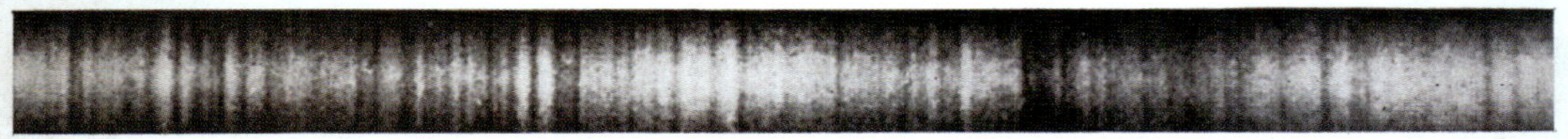

World without Neighbors

This full-color photograph of the planet Saturn and its fabulous rings was made on Kodachrome Film at the Mt. Wilson Observatory.

Weird and beautiful though it is—and a great achievement in photography—it has less significance than the black-and-white strip, also photographed at Mt. Wilson on a special Kodak plate sensitive to wave lengths longer than those of light. For the spectroscopic strip informs the scientist that Saturn, with 81 times the "living space" of the Earth, is a world without men and women—a world without any form of life as we know it . . .

The analyst finds in Saturn's spectrum evidence of ammonia gas and methane gas in large quantities. He finds no free oxygen or carbon dioxide —essential to good or bad neighbors, of the kind we know, and to plant life.

The branch of photography known as *spectroscopy* has told us all we know about the composition and atmosphere of the stars. The practical man may wave that aside as "interesting, but what of it?".

Yet it is through this same photographic process that the practical man in plant or laboratory analyzes metals, or other compounds, in minutes instead of hours—maintaining quality, or leading to improvement, of the products you buy.

EASTMAN KODAK COMPANY, Rochester 4, N. Y.

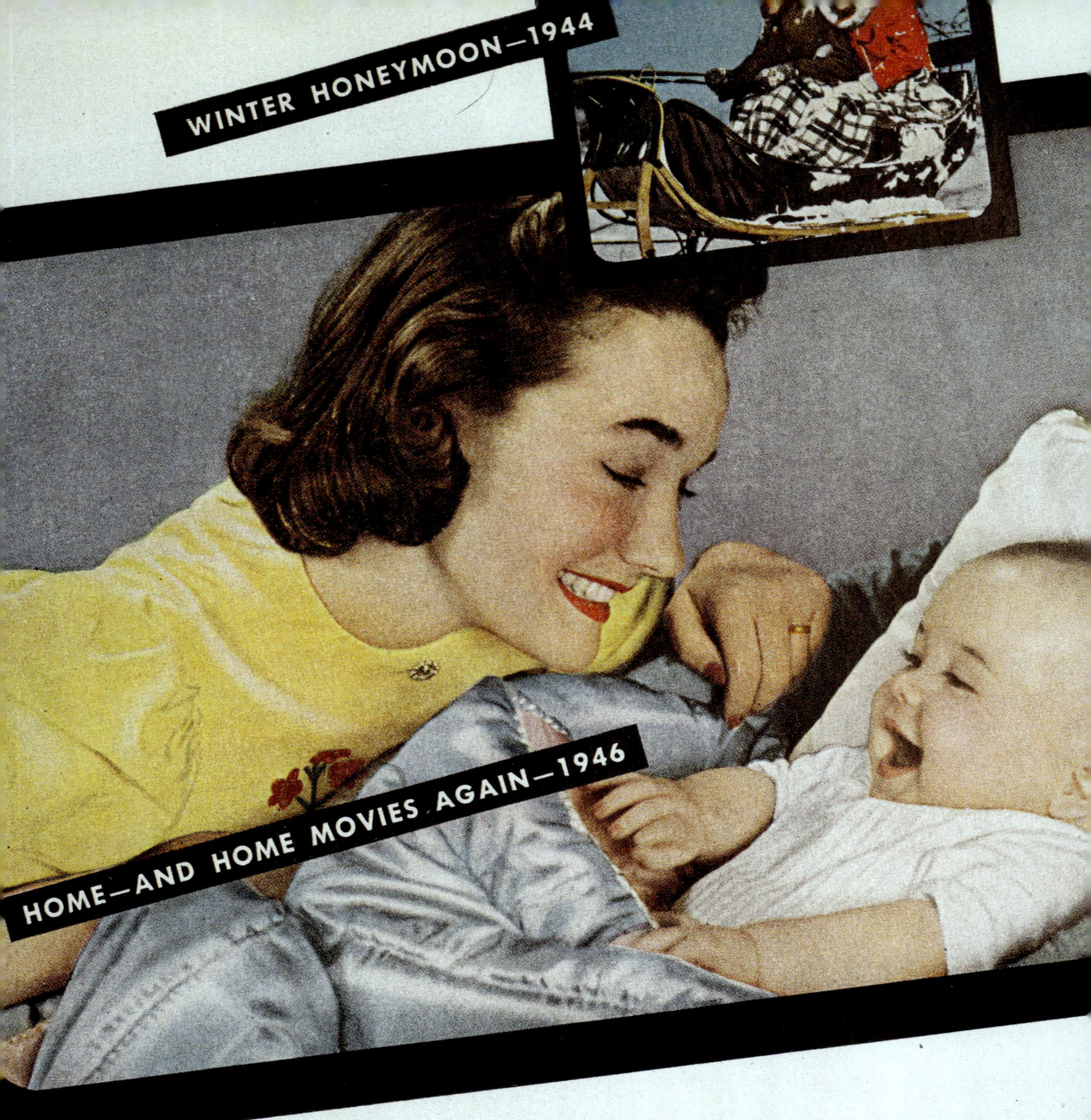

WINTER HONEYMOON—1944
HOME—AND HOME MOVIES AGAIN—1946
Make up for lost time...and lost m
start your NEW movie record now
Ciné-Kodaks are on the way!
No more film

The Post-War Era
1946-1979

Hey soldier! Remember that quick vacation you and your new bride took back in the Winter of '45? Well, the results are in—it's a boy and it's time to start making memories.

The war came to an end in 1945, and consumer spending was on the uptick. Our soldiers were returning home, and the Baby Boomers began making their grand debut. Life was good again, and no film was better at capturing the colors of life than Kodachrome.

The focus of Kodak's post-war advertising campaign was a simple one: building memories. After years of war, life's celebrations were once again front and center, and what better way of preserving memories than true-to-life color photographs.

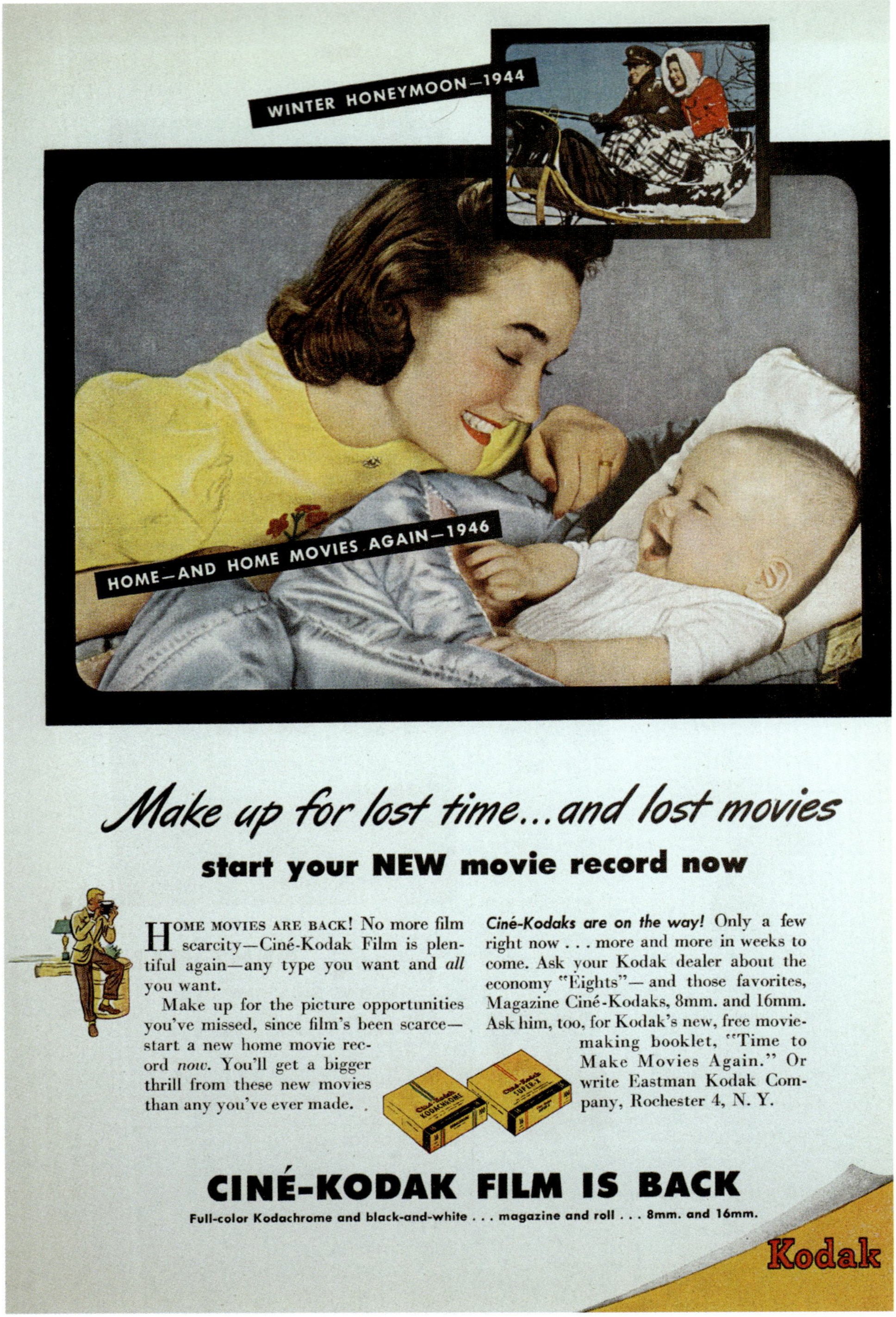

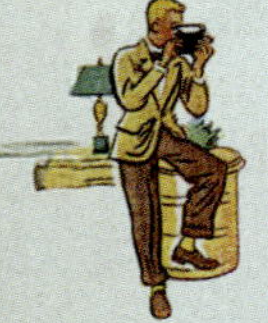

WINTER HONEYMOON—1944

HOME—AND HOME MOVIES AGAIN—1946

Make up for lost time...and lost movies

start your NEW movie record now

HOME MOVIES ARE BACK! No more film scarcity—Ciné-Kodak Film is plentiful again—any type you want and *all* you want.

Make up for the picture opportunities you've missed, since film's been scarce—start a new home movie record *now*. You'll get a bigger thrill from these new movies than any you've ever made.

Ciné-Kodaks are on the way! Only a few right now . . . more and more in weeks to come. Ask your Kodak dealer about the economy "Eights"— and those favorites, Magazine Ciné-Kodaks, 8mm. and 16mm. Ask him, too, for Kodak's new, free movie-making booklet, "Time to Make Movies Again." Or write Eastman Kodak Company, Rochester 4, N. Y.

CINÉ-KODAK FILM IS BACK

Full-color Kodachrome and black-and-white . . . magazine and roll . . . 8mm. and 16mm.

Kodak

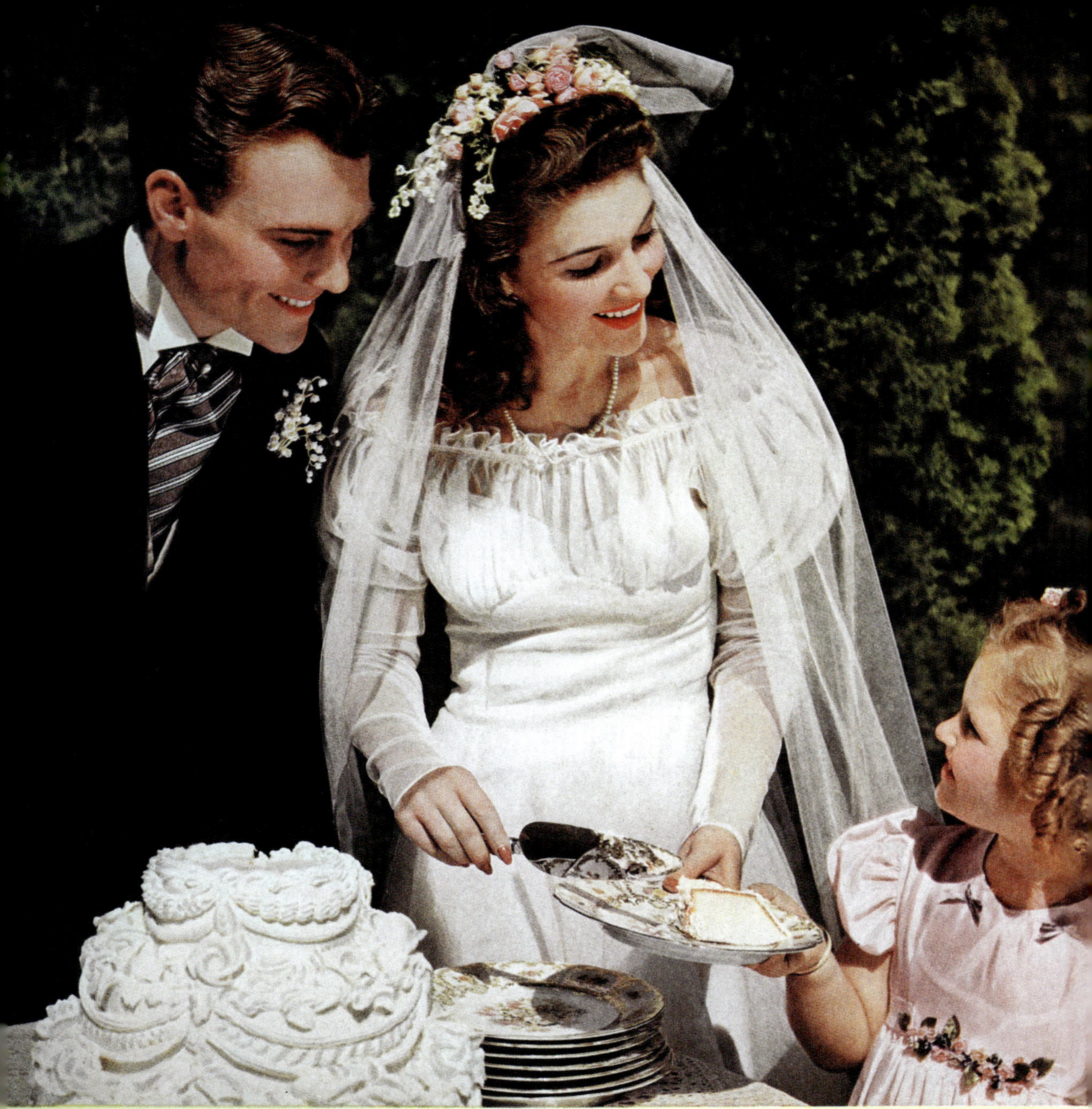

You can take color pictures as lovely and as memorable as this one, if you make home movies on full-color Kodachrome Film. Every home movie camera that Eastman makes, and that means all Ciné-Kodaks, loads with this wonderful color film. See your Ciné-Kodak dealer ... Eastman Kodak Company, Rochester, N.Y.

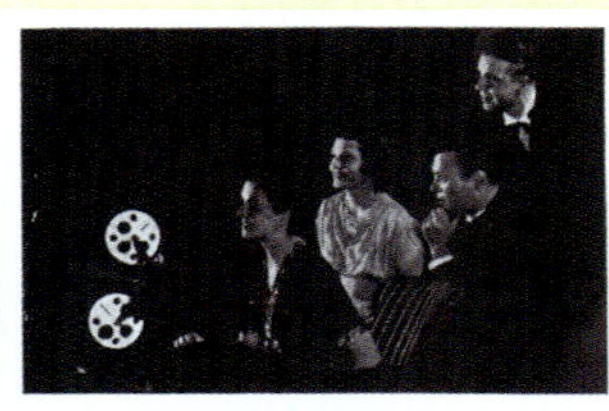

Kodachrome Film

EASTMAN'S FULL-COLOR
HOME MOVIE FILM

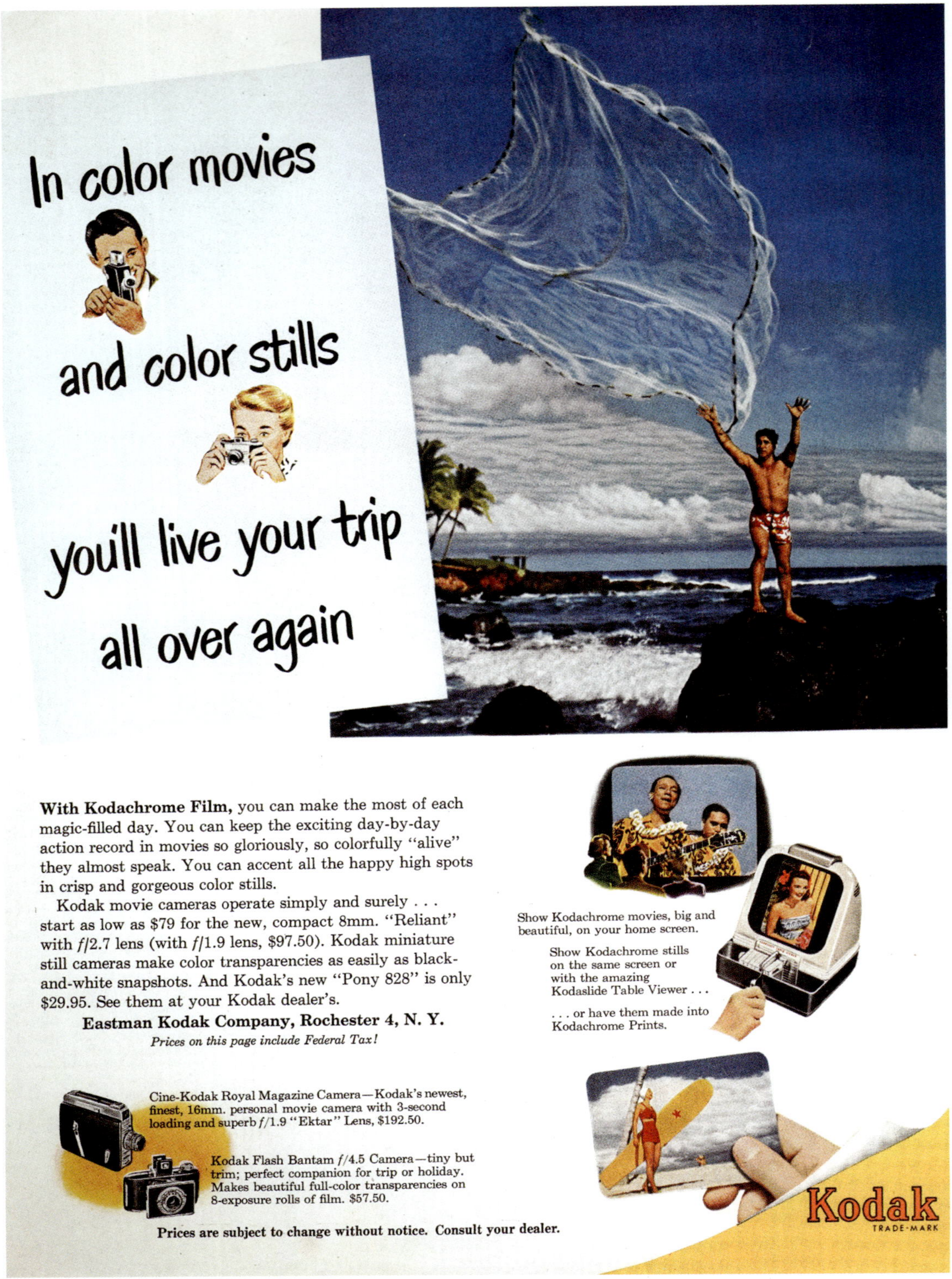

With Kodachrome Film, you can make the most of each magic-filled day. You can keep the exciting day-by-day action record in movies so gloriously, so colorfully "alive" they almost speak. You can accent all the happy high spots in crisp and gorgeous color stills.

Kodak movie cameras operate simply and surely . . . start as low as $79 for the new, compact 8mm. "Reliant" with $f/2.7$ lens (with $f/1.9$ lens, $97.50). Kodak miniature still cameras make color transparencies as easily as black-and-white snapshots. And Kodak's new "Pony 828" is only $29.95. See them at your Kodak dealer's.

Eastman Kodak Company, Rochester 4, N. Y.

Prices on this page include Federal Tax!

Cine-Kodak Royal Magazine Camera—Kodak's newest, finest, 16mm. personal movie camera with 3-second loading and superb $f/1.9$ "Ektar" Lens, $192.50.

Kodak Flash Bantam $f/4.5$ Camera—tiny but trim; perfect companion for trip or holiday. Makes beautiful full-color transparencies on 8-exposure rolls of film. $57.50.

Prices are subject to change without notice. Consult your dealer.

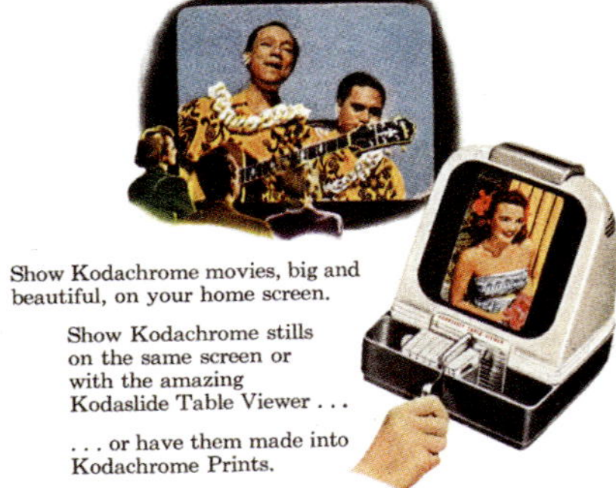

Show Kodachrome movies, big and beautiful, on your home screen.

Show Kodachrome stills on the same screen or with the amazing Kodaslide Table Viewer . . .

. . . or have them made into Kodachrome Prints.

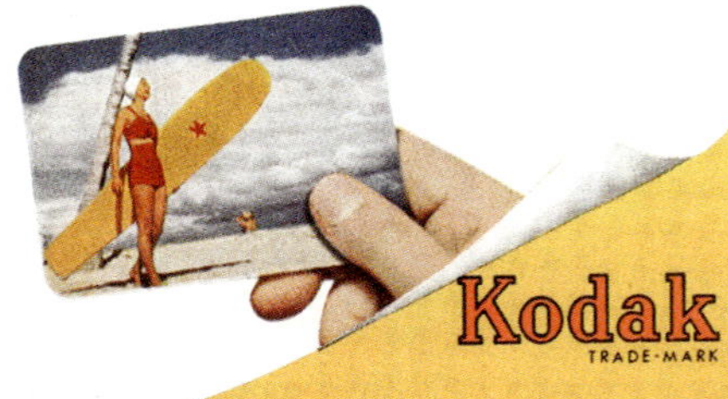

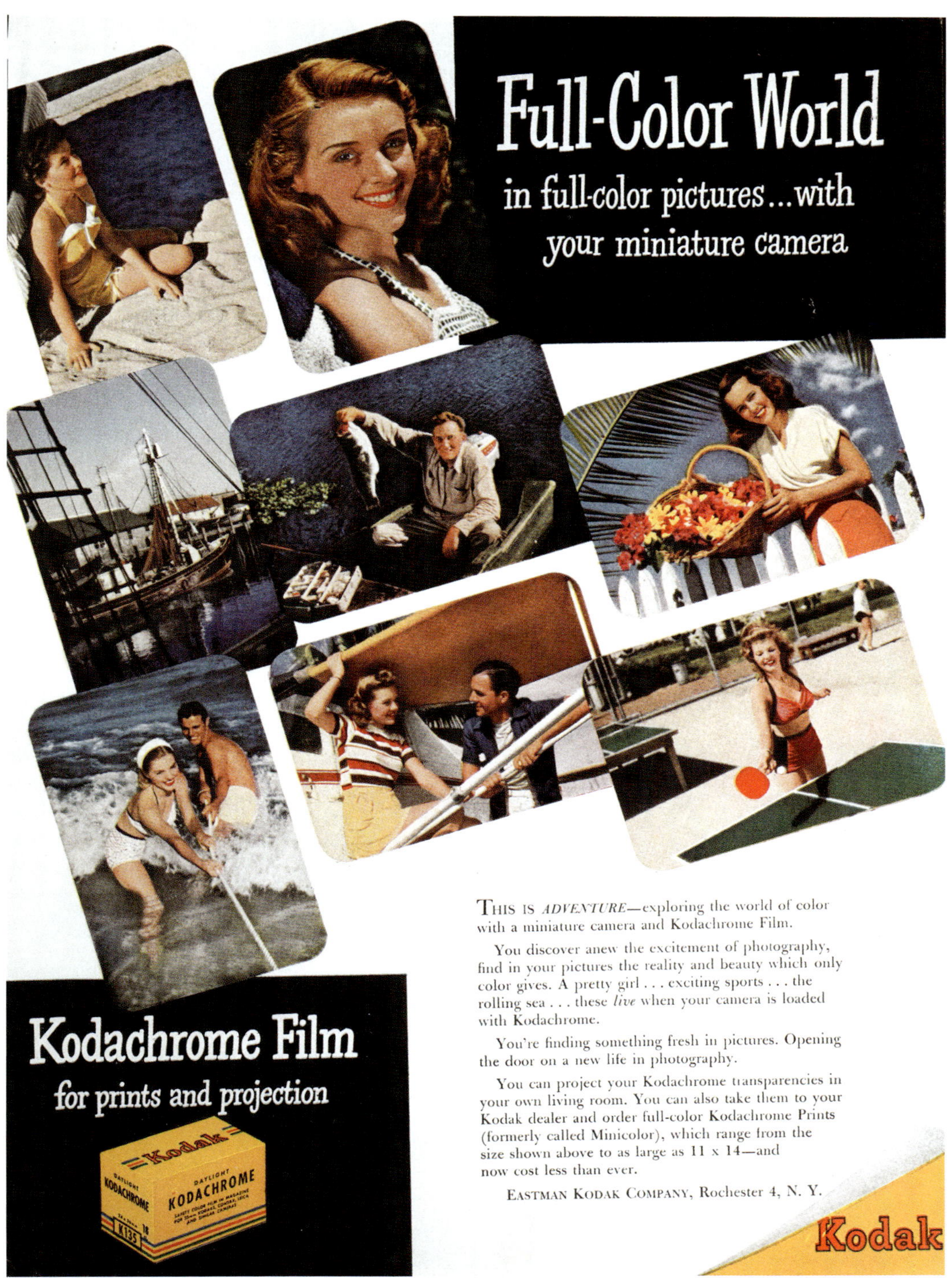

◀ 1947, USA   ▲ 1948, USA   ▶▶ 1947, USA

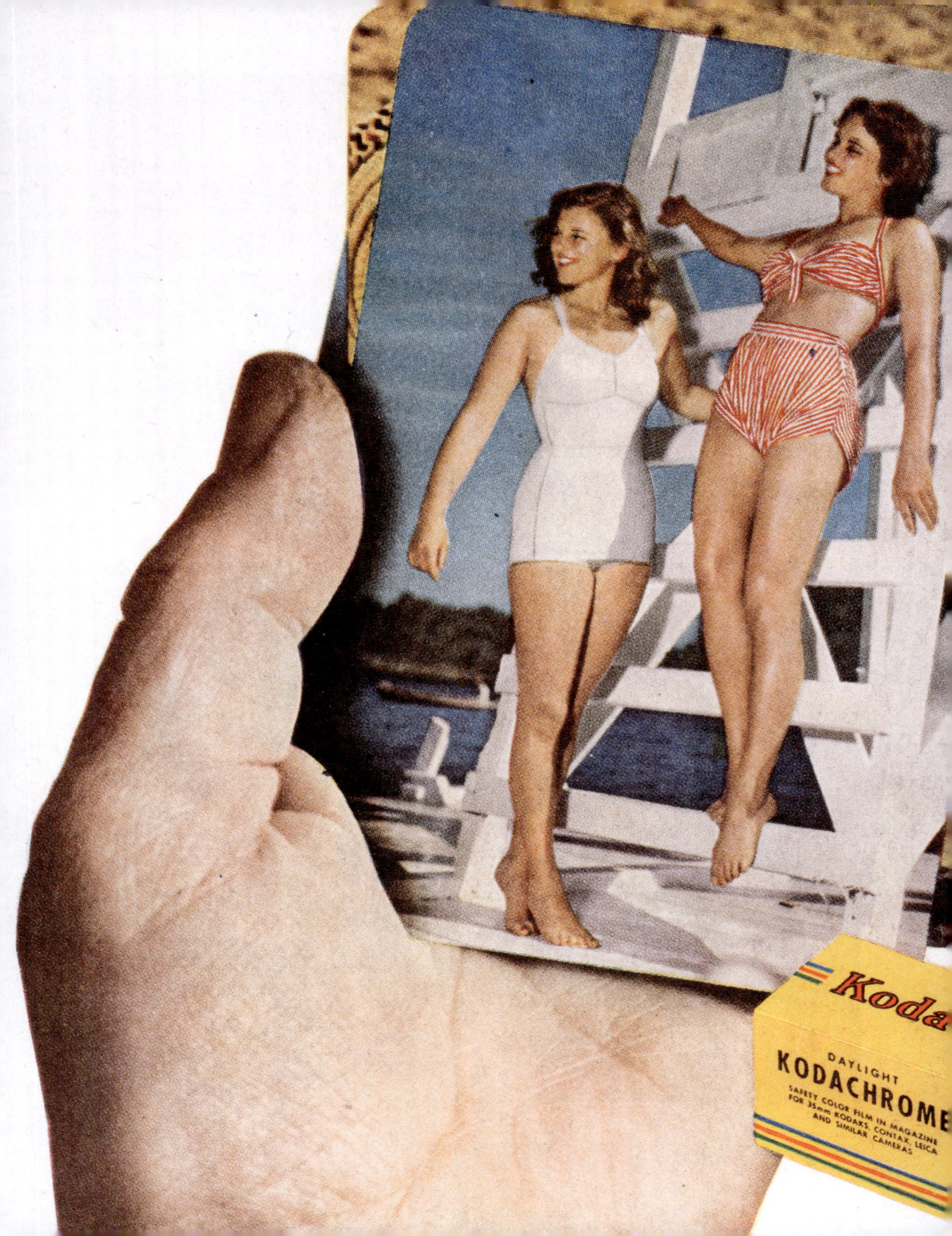
Koda
DAYLIGHT
KODACHROME
SAFETY COLOR FILM IN MAGAZINE
FOR 35mm KODAKS, CONTAX, LEICA
AND SIMILAR CAMERAS

Prints: Your Kodachrome Prints are
fine examples of full-color printing.

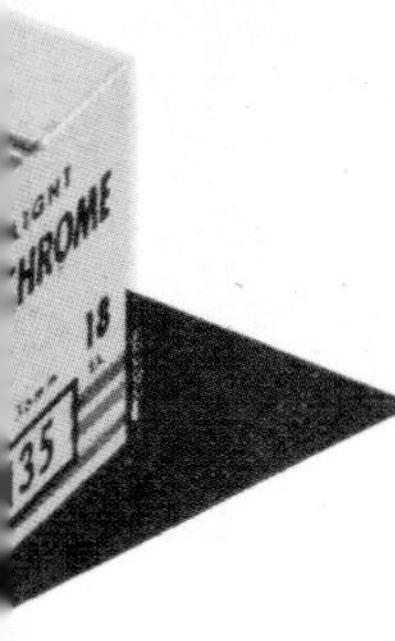

SUMMER remains "very truly yours" when you load your miniature camera with Kodachrome Film . . .

For summer *is* color. How breath-takingly evident this becomes when your exposed film is returned through your Kodak dealer, finished without charge in the form of Kodachrome transparencies . . . and you project them on your screen with a Kodaslide projector!

Your second reward comes when you order Kodachrome Prints—and see how your friends respond to these full-color snapshots. Order the new, reasonably priced 3X size shown here . . . or larger or smaller sizes . . . through your dealer.

Start your color career—get Kodachrome *now* . . .

EASTMAN KODAK COMPANY
ROCHESTER 4, N. Y.

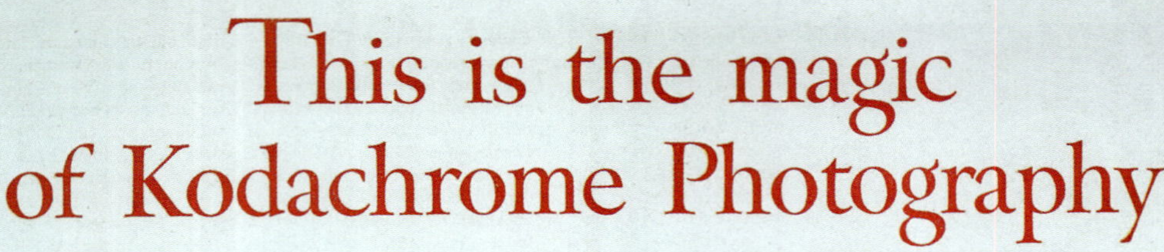

Some wonderful day you'll load a **miniature camera** with **Kodachrome Film** and shoot. When your pictures come back they're **color slides** ... breathtakingly beautiful when you project them on a **home screen.** And you can have sparkling **color prints** made from them, too. This is the magic of Kodachrome photography.

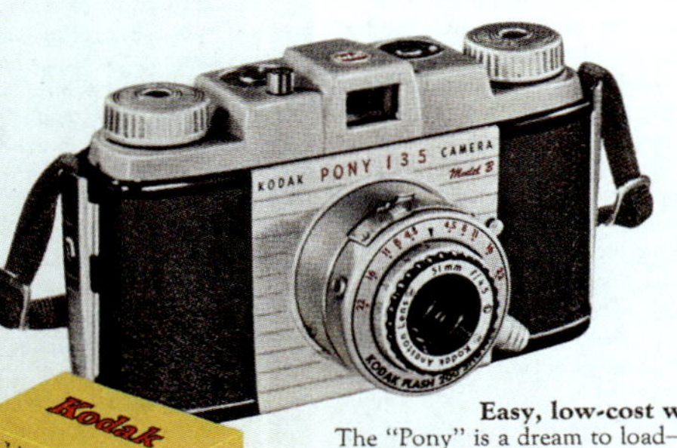

Easy, low-cost way to Kodachrome pictures
The "Pony" is a dream to load—no film threading required. Average settings marked in red give you box-camera simplicity. Automatic film stop, film count. And, of course, the fine, fast f/4.5 lens means crisp, clear color pictures indoors as well as out. All this at a surprisingly moderate price.

Kodak Pony 135 Camera, Model B, $34.75
Flasholder, $7.95

Many Kodak dealers offer convenient terms

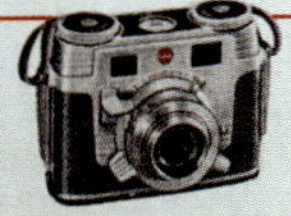

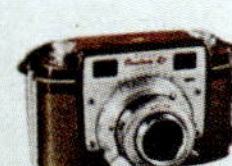

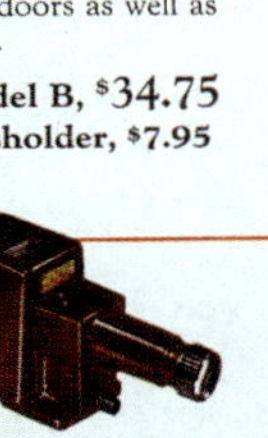

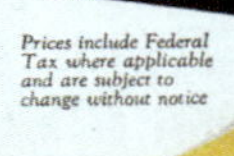

Finest medium-priced color camera·... Kodak Signet 35 Camera with its unrivalled Ektar f/3.5 Lens for extra color purity, extra brilliance, extra detail. Coupled rangefinder and automatic controls, of course. $87.50. Flasholder, $7.95.

Kodak's newest color camera ... the "Bantam RF," with rangefinder, f/3.9 lens, automatic controls. Handsome two-toned styling—a camera you'll be proud to own. In attractive outfit, with Flasholder and case, $75. Camera only, $59.75.

To show your slides ... Kodak offers a full range of fine projectors starting with the "Merit" shown above. Features new-type slide feeding at top to eliminate jarring. With 150-watt lamp and 10-foot cord, only $23.95.

Prices include Federal Tax where applicable and are subject to change without notice

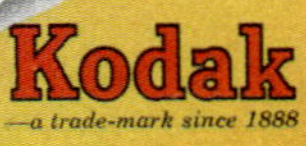

Eastman Kodak Company, Rochester 4, N. Y.

Kodachrome was more than a color film.
It was the starting point for selling Kodak
Pony, Retina, and Bantam still cameras,
slide projectors, Cine-Kodak 8mm movie
cameras, projection screens, color prints,
and related accessories.

For decades it was possible to fill all of
your photographic needs, i.e., still cameras,
movie cameras, film, and everything you
would need to view and project your still
and movie imagery without having to leave
the fold. Kodak had you covered from
soup to nuts with photo gear designed
to fit every budget. And the best part? It
was all top quality.

◀ 1954, USA

▶ 1959, USA

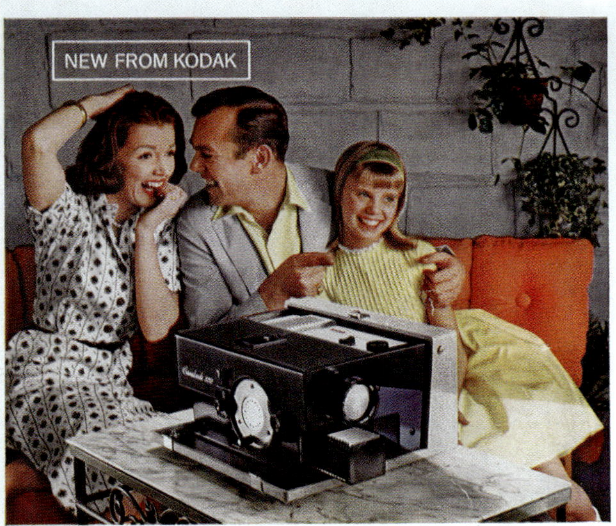

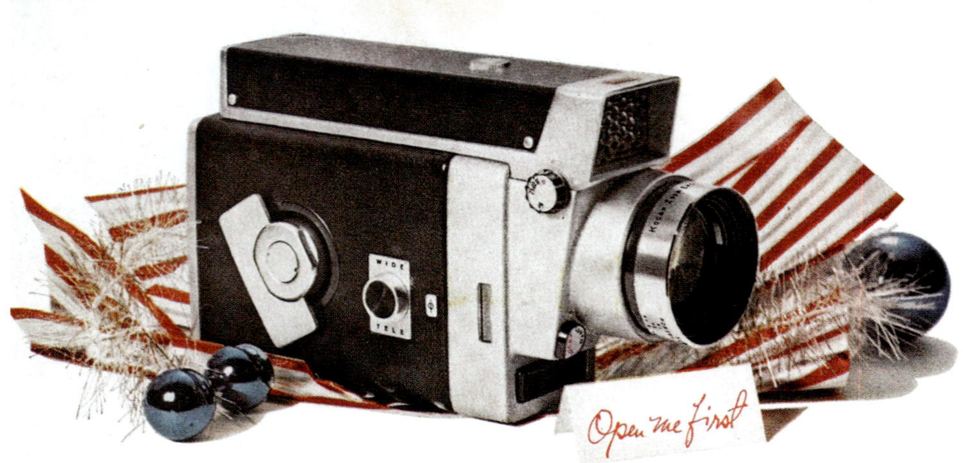

Push-button zooming! Fully automatic! This de luxe 8mm *Kodak Zoom 8 Reflex Camera* lets you make zoom shots with push-button ease. You view through the lens. Electric eye sets exposure. Less than $200.

Kodak gifts say "Open me first"
...and <u>picture</u> all your Christmas fun!

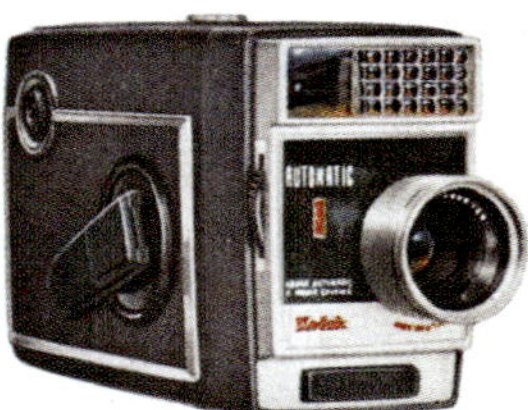

New automatic 8mm movie camera—budget price! *Kodak Automatic 8 Movie Camera* sets its own lens and has filter built in. Less than $50.

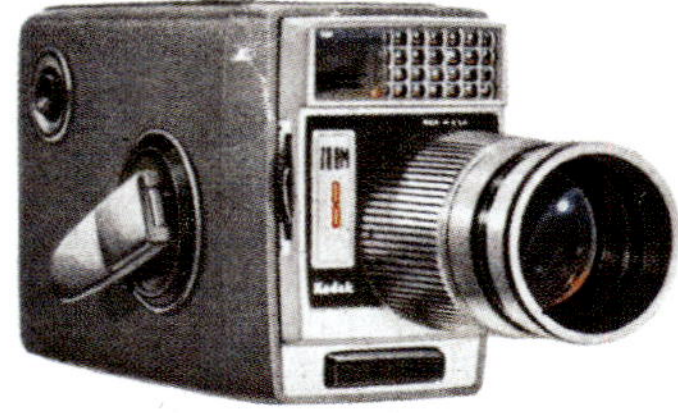

Get dramatic 8mm movie zoom shots with the new *Kodak Zoom 8 Automatic Camera.* Has electric eye, built-in filter. Less than $110.

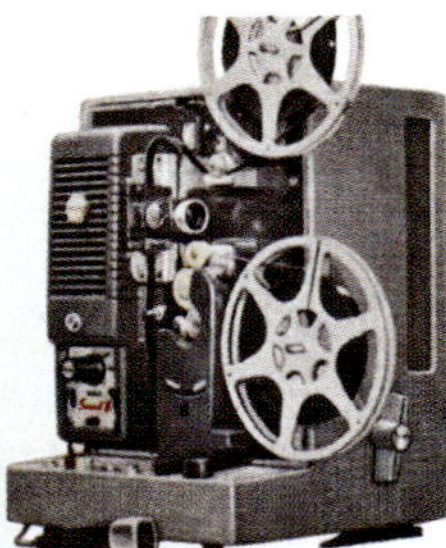

Add quality sound to your movies! *Kodak Sound 8 Projector* lets you add your own commentary, music and sound effects to 8mm films. Less than $350.

New! *Kodachrome II Film* is 2½ times as fast, gives you better color, greater sharpness, and wider exposure latitude. Here's an ideal stocking filler for movie fans.

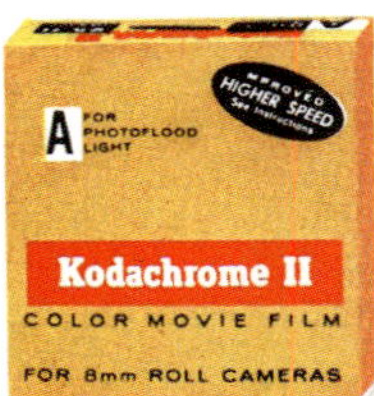

Prices subject to change without notice.

Kodak
TRADEMARK

EASTMAN KODAK COMPANY, Rochester 4, N.Y.

1961, USA ▲

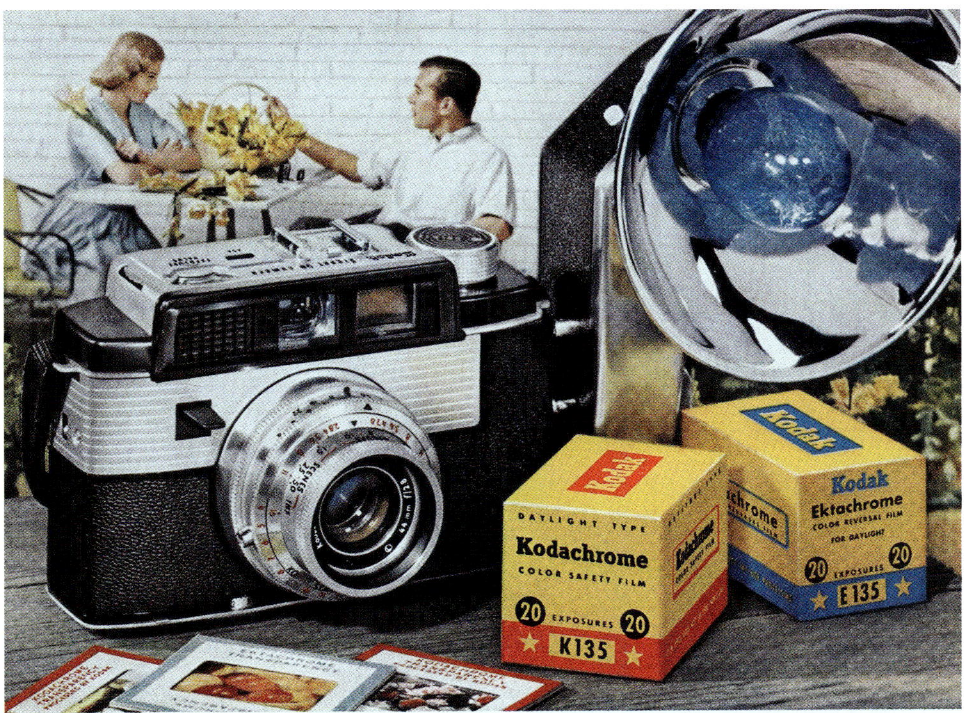

A superb new 35mm camera from the complete new Kodak selection

New Kodak Signet 50 Camera has light meter built in...gives you gorgeous color slides!

Whether you've taken color slides by the hundreds, or have yet to make your first one—you'll be a better photographer with the new Kodak Signet 50 Camera.

This fine 35mm camera takes the guesswork out of exposure with a built-in photoelectric meter. It has the new, simplified Light Value System . . . a fast, superbly sharp *f*/2.8 lens . . . and single-stroke film advance lever. Ask your photo dealer to demonstrate this new camera. With flasholder and two reflectors, it's only $82.50, or as little as $8.50 down. Also see the Kodak Signet 30—the same camera, without exposure meter and flash equipment—for $55. (Prices are list, include Federal Tax and are subject to change without notice.)

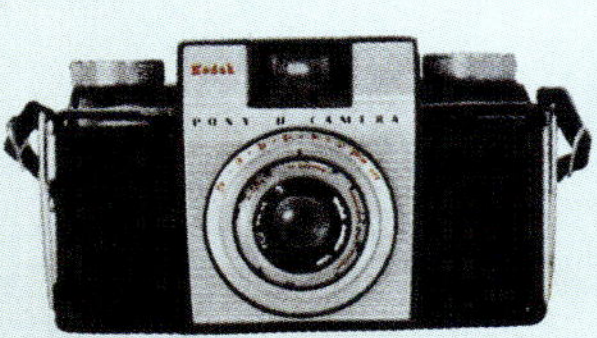

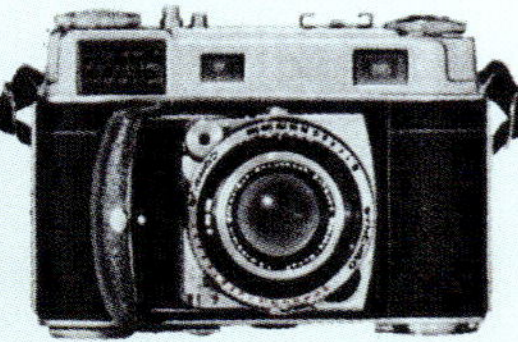

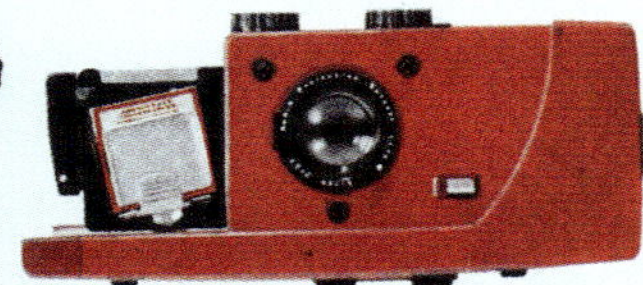

New Kodak Pony II Camera... $26.75—Easiest way to advance to Kodak 35mm color slides! Just two settings to make. $3 down.

New Kodak Retina IIIc Camera ...$175—World-famous . . . has rangefinder, built-in light meter, *f*/2 lens. $17.50 down.

New Kodak 300 Projector . . . $64.50—Shows color slides big and bright; has new Ready-matic Changer. $6.50 down.

See Kodak's TV shows—"The Ed Sullivan Show" and "The Adventures of Ozzie and Harriet."

EASTMAN KODAK COMPANY, Rochester 4, N.Y.

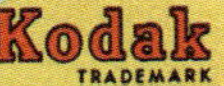

1958, USA

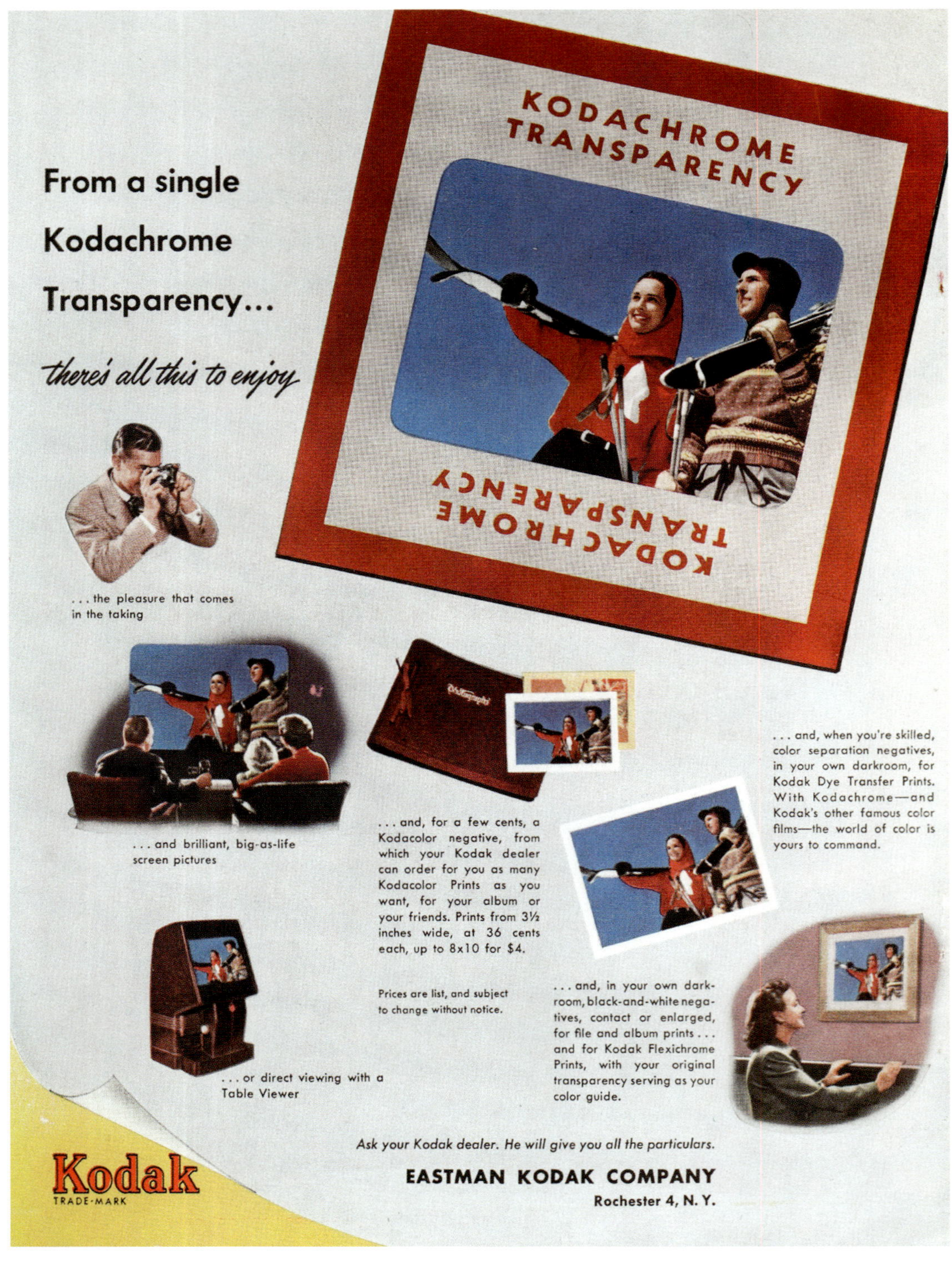

1956, USA ▲   1954, USA ▶   1954, USA ▶▶

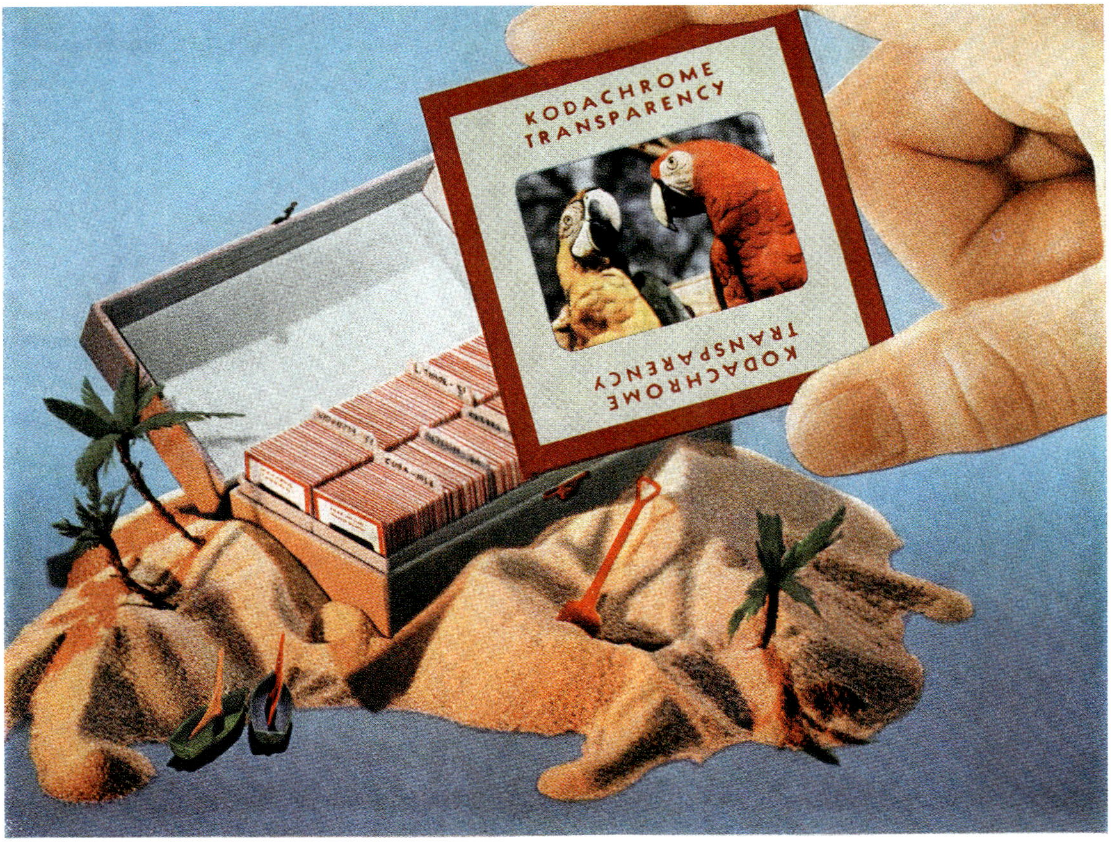

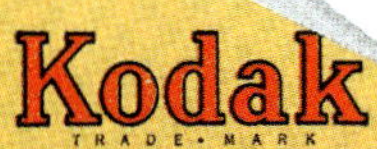

Treasure Island...

Blackbeard and Captain Kidd had a poor deal. They had to spend their booty—or worse, bury it and lose it forever. They took big risks—for a slim reward.

Buccaneering with Kodachrome Film is far better. You find your own treasure . . . raid the world's color, yet leave it there, unhurt. You bring it home . . . enjoy it . . . share it . . . and the more you share, the more you keep.

There's adventure in Kodachrome—and a rich reward in sparkling color transparencies. Thrift, too—only $1.88 for 8 shots in 828 film, and in 135 film only $3.50 for 20 shots, $5.50 for 36—all processed, mounted as slides, and mailed to you. It's the quick, happy route to a Treasure Island of your own.

That treasure chest is a Kodaslide Flexo File. Only $1.25, it will hold 360 colorful memories—or 160 in stereo. At your Kodak dealer's.

Prices include Federal Tax and are subject to change without notice.

EASTMAN KODAK COMPANY, Rochester 4, N.Y.

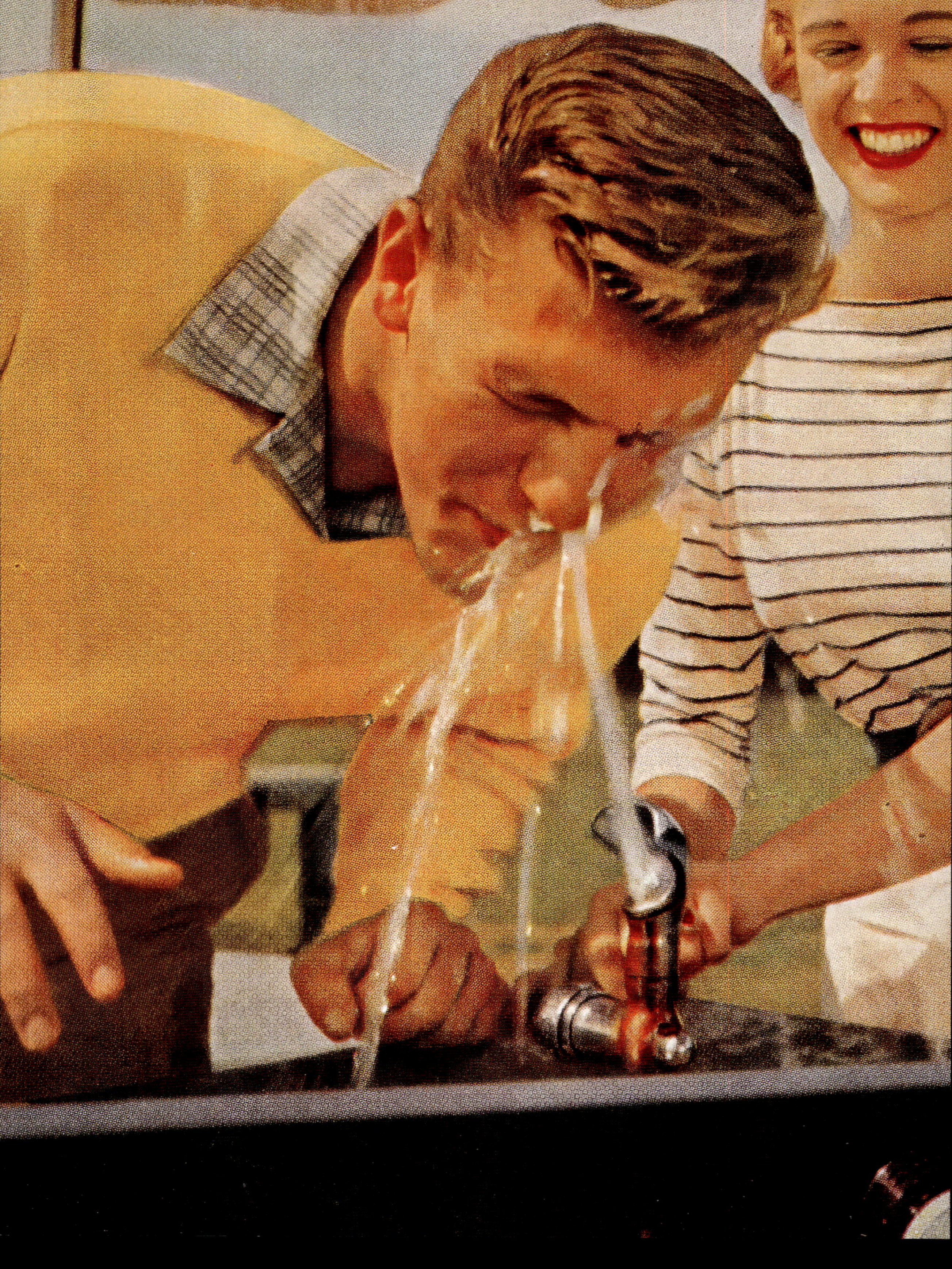

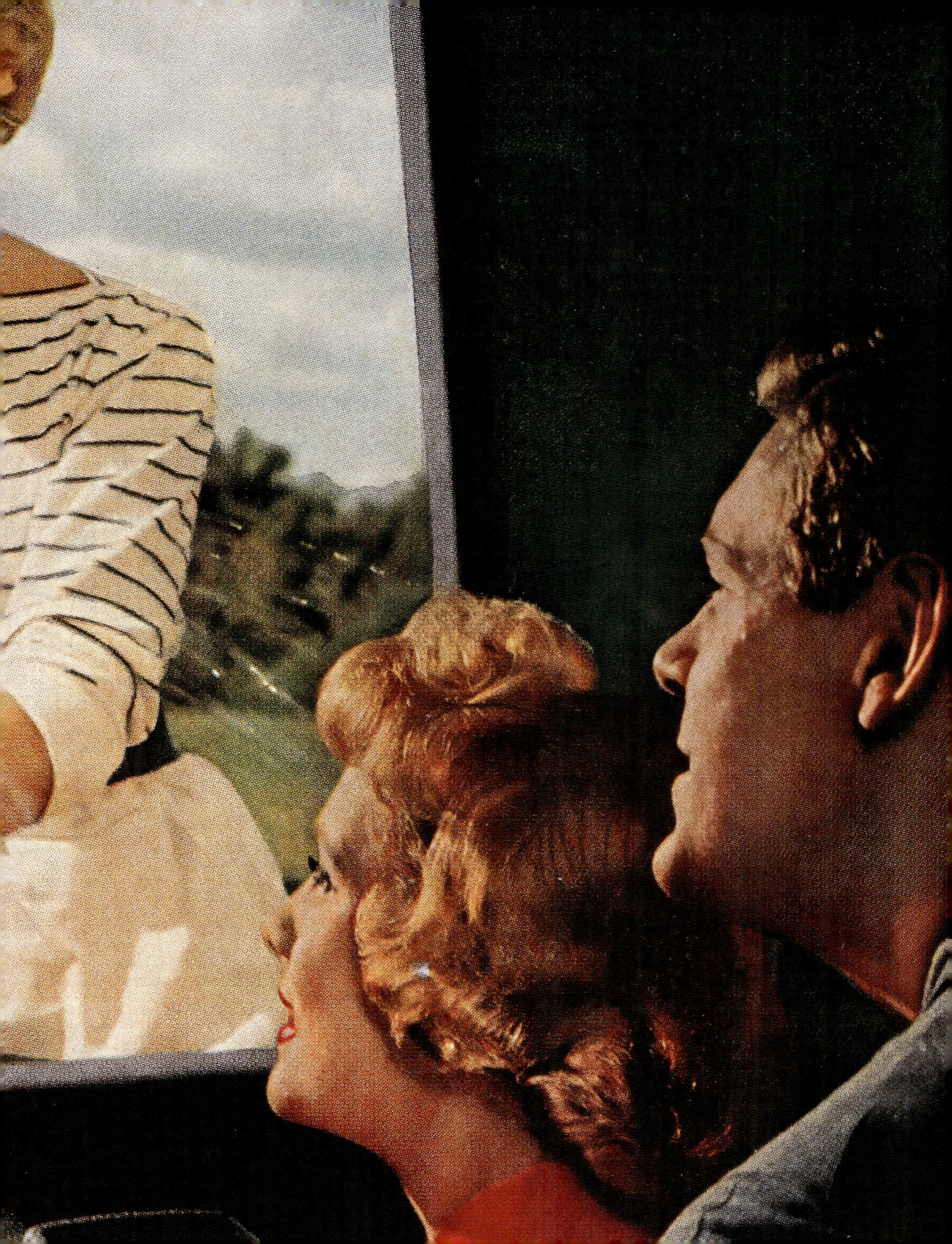

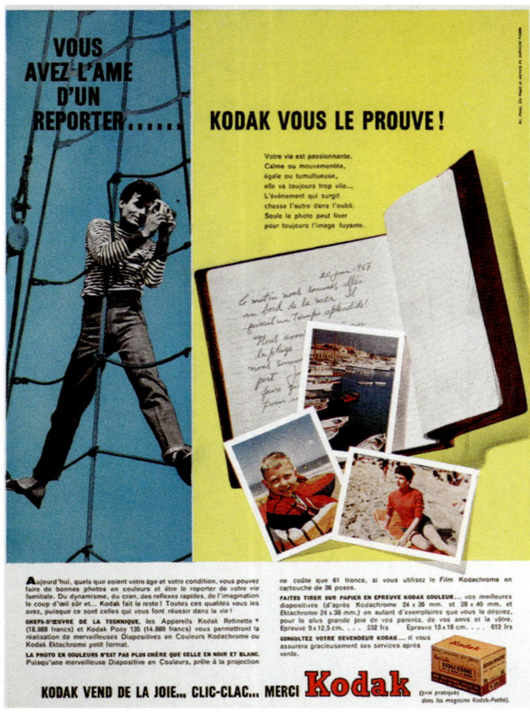

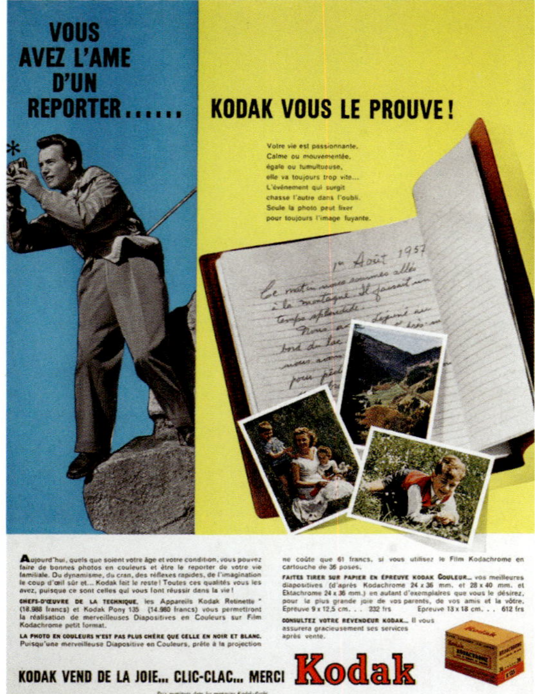

No matter which side of the Atlantic you lived on, the message was the same—when it comes to saving memories, no film does it better than Kodachrome!

1957, France ▲

1956, France ▶

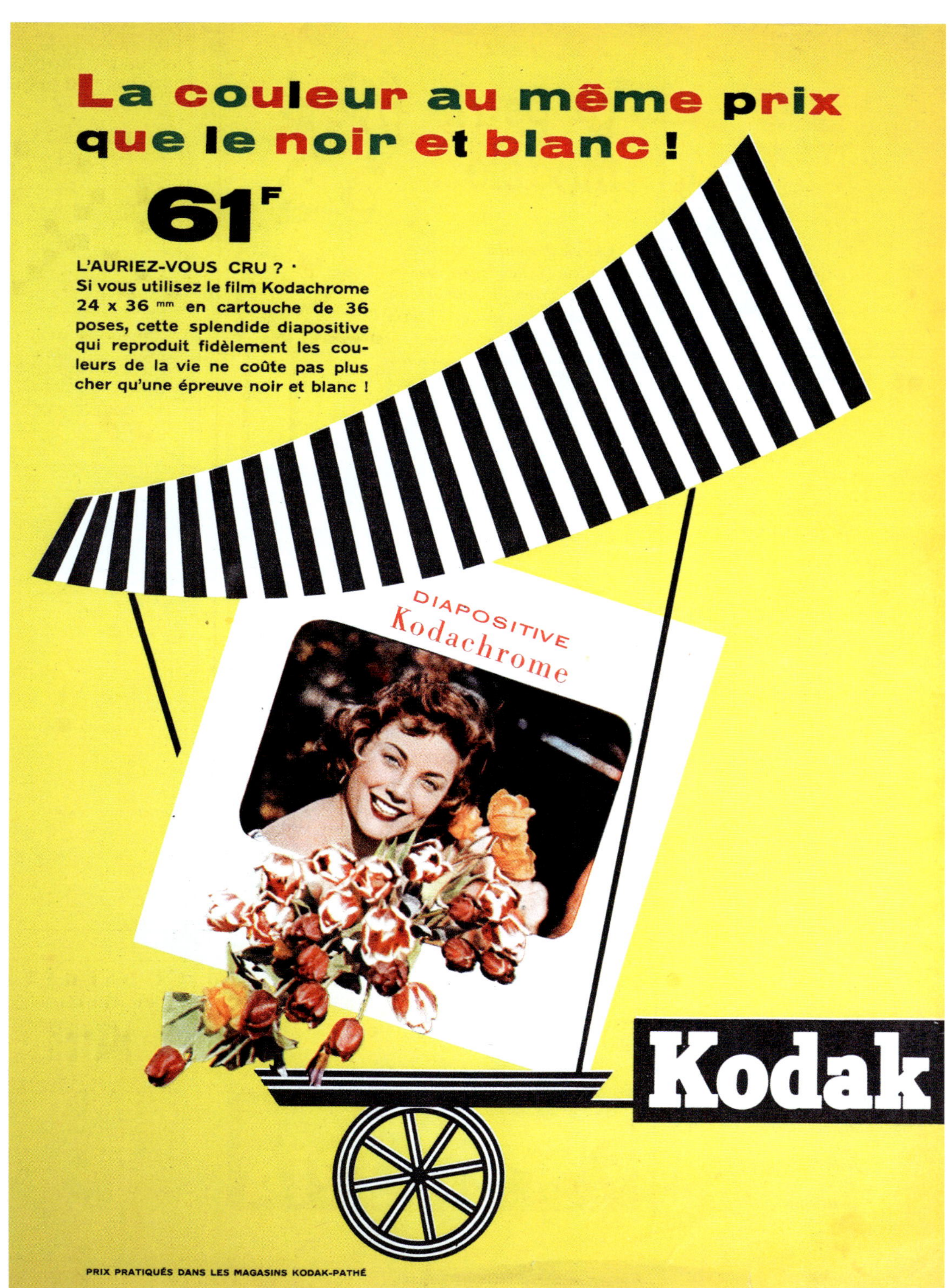

La couleur au même prix que le noir et blanc !
61F
L'AURIEZ-VOUS CRU ? ·
Si vous utilisez le film Kodachrome 24 x 36 mm en cartouche de 36 poses, cette splendide diapositive qui reproduit fidèlement les couleurs de la vie ne coûte pas plus cher qu'une épreuve noir et blanc !
DIAPOSITIVE Kodachrome
Kodak
PRIX PRATIQUÉS DANS LES MAGASINS KODAK-PATHÉ

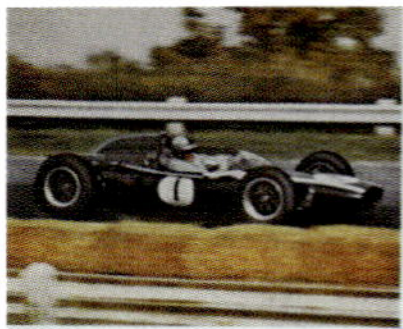
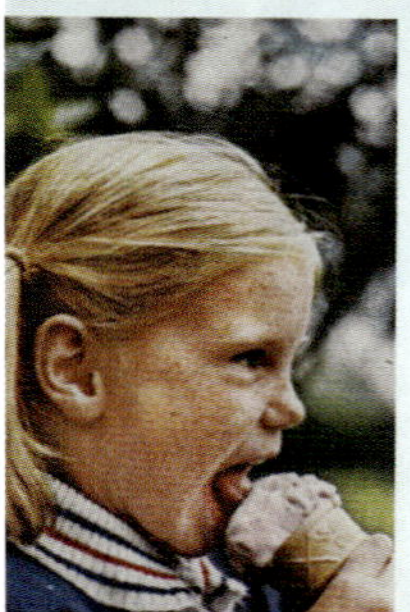
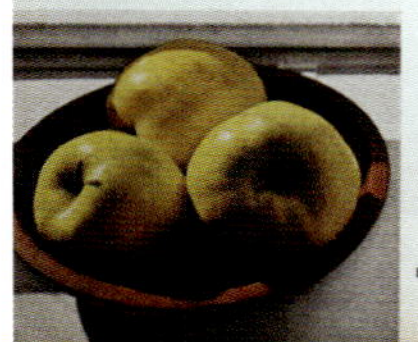
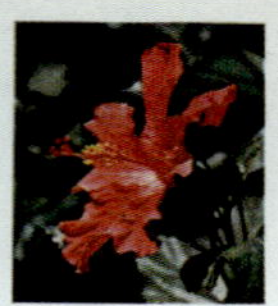
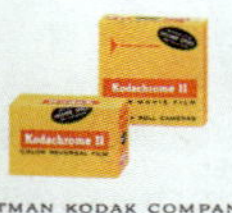
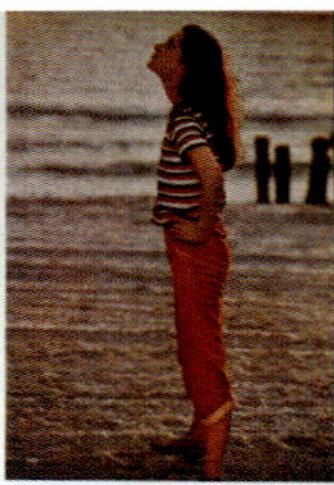
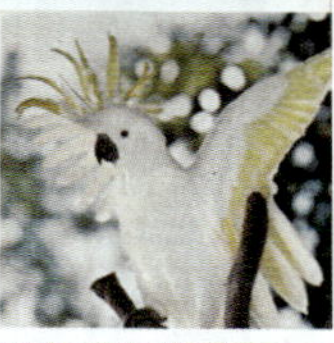
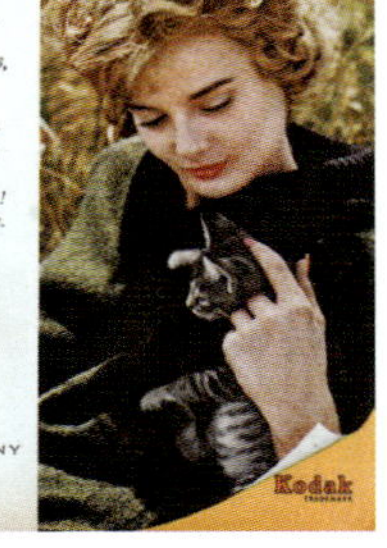

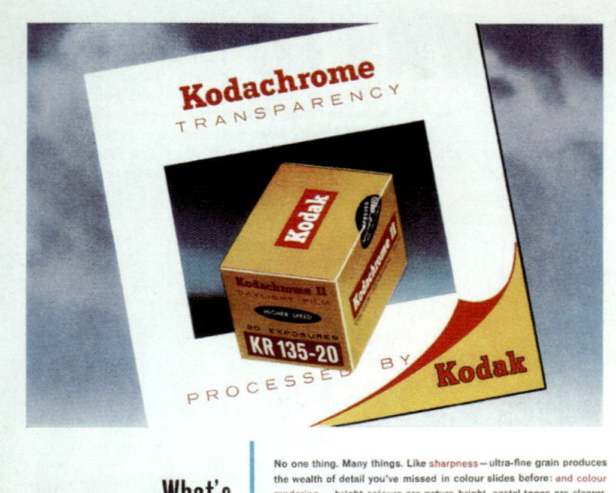

1962, USA ▲

1963, USA ▲

1963, UK ▶

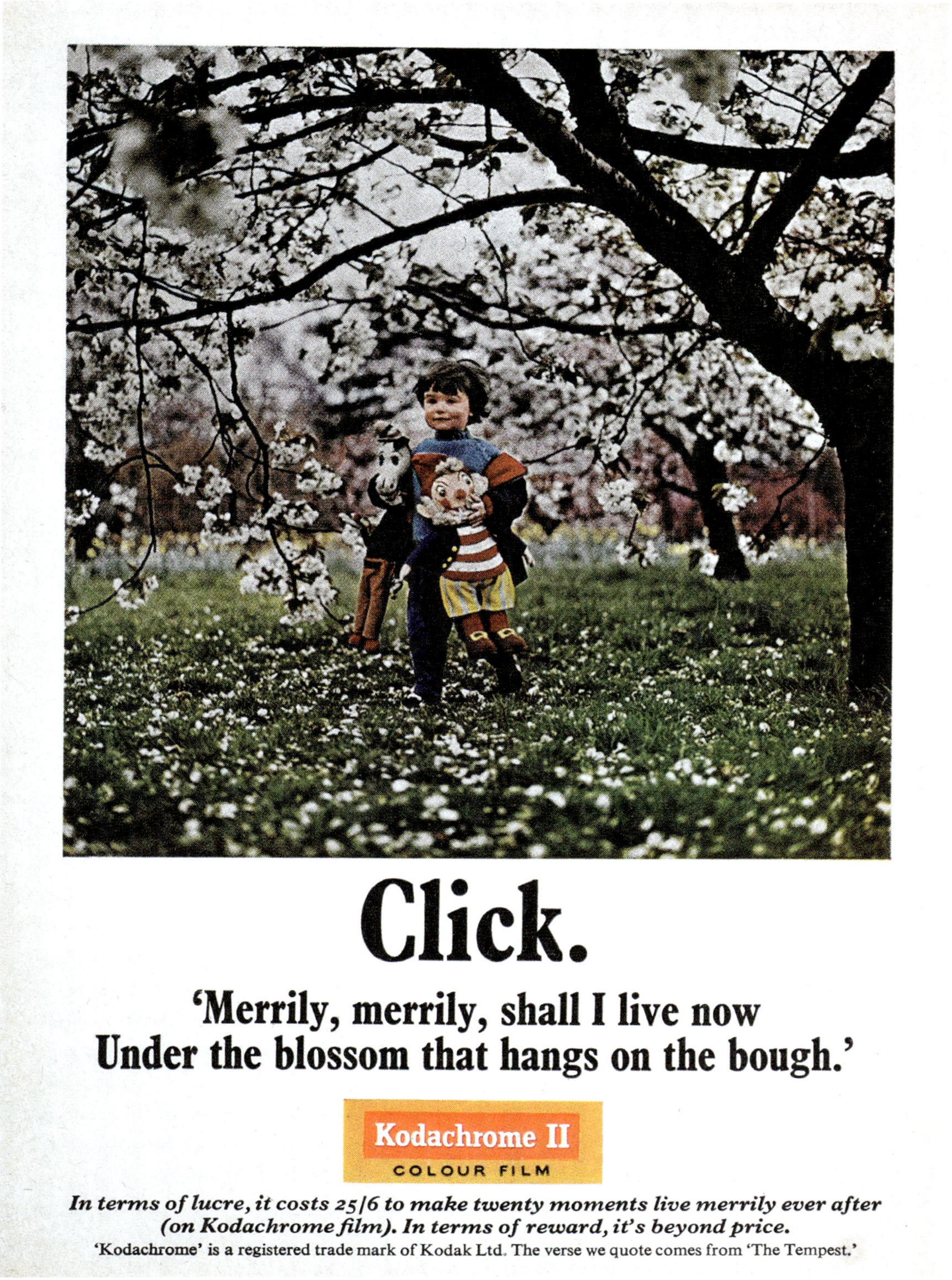

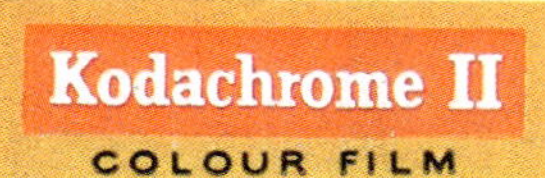

Click.

**'Merrily, merrily, shall I live now
Under the blossom that hangs on the bough.'**

*In terms of lucre, it costs 25/6 to make twenty moments live merrily ever after
(on Kodachrome film). In terms of reward, it's beyond price.*

'Kodachrome' is a registered trade mark of Kodak Ltd. The verse we quote comes from 'The Tempest.'

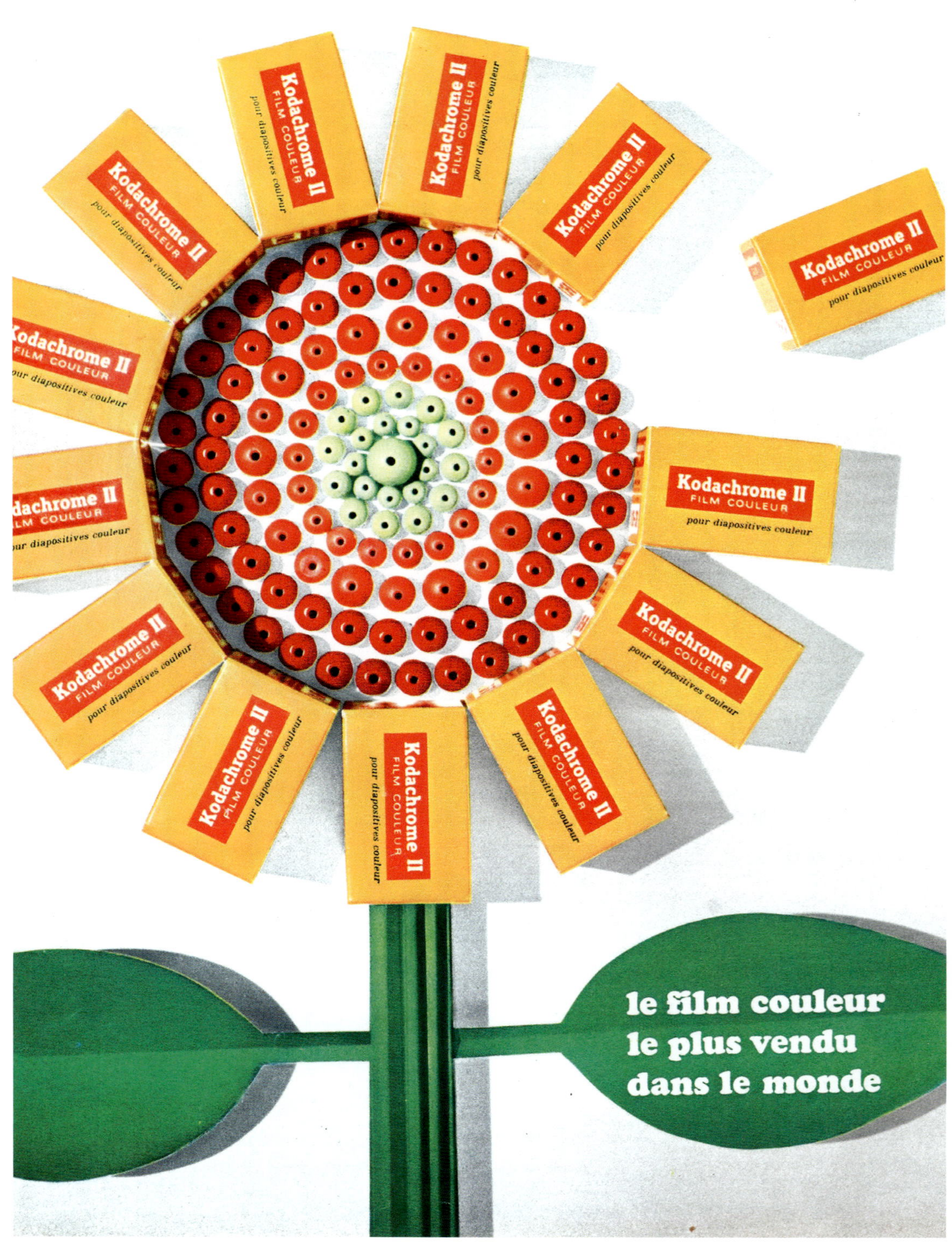

Kodachrome II
FILM COULEUR
pour diapositives couleur
le film couleur
le plus vendu
dans le monde

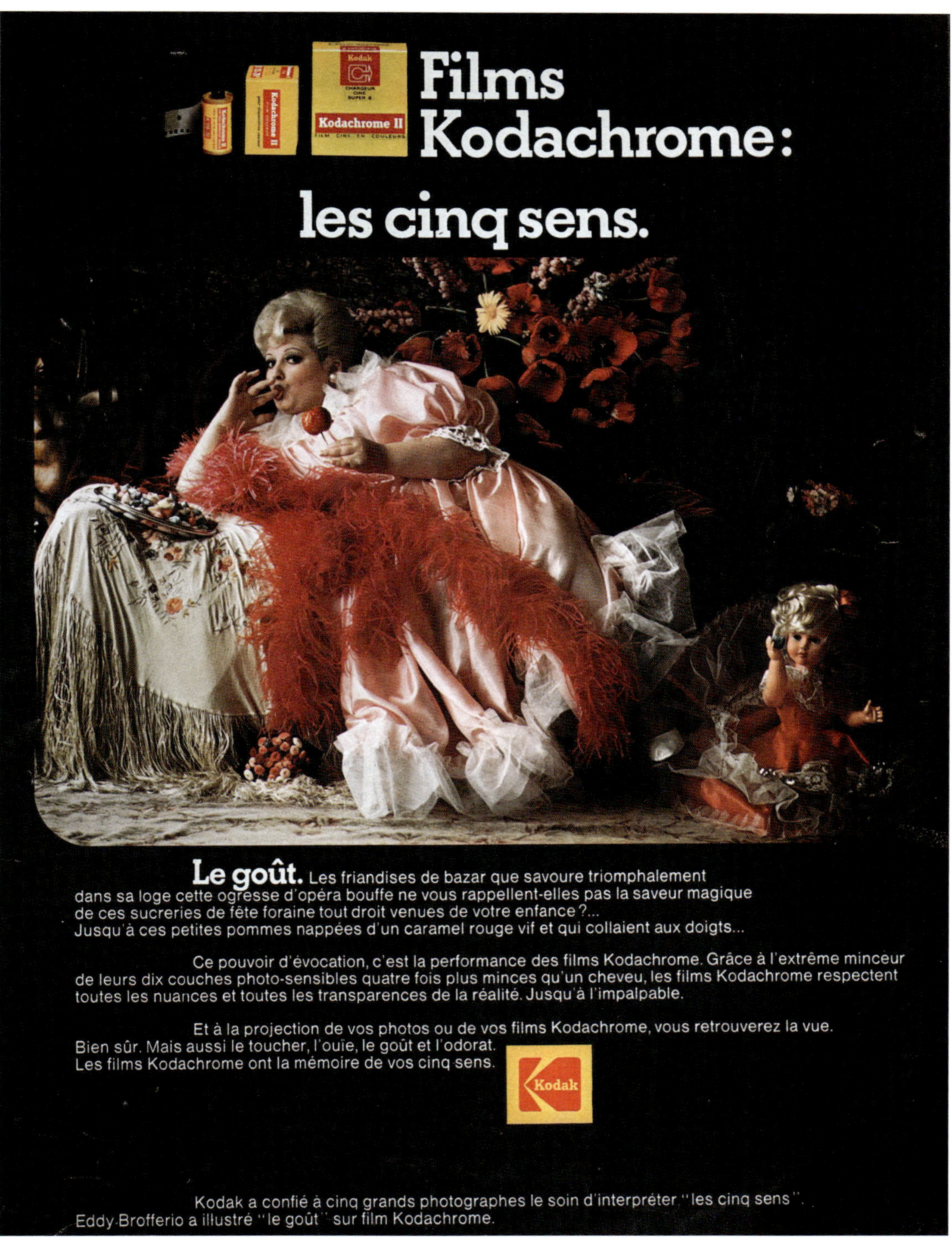

1967, France   1972, France

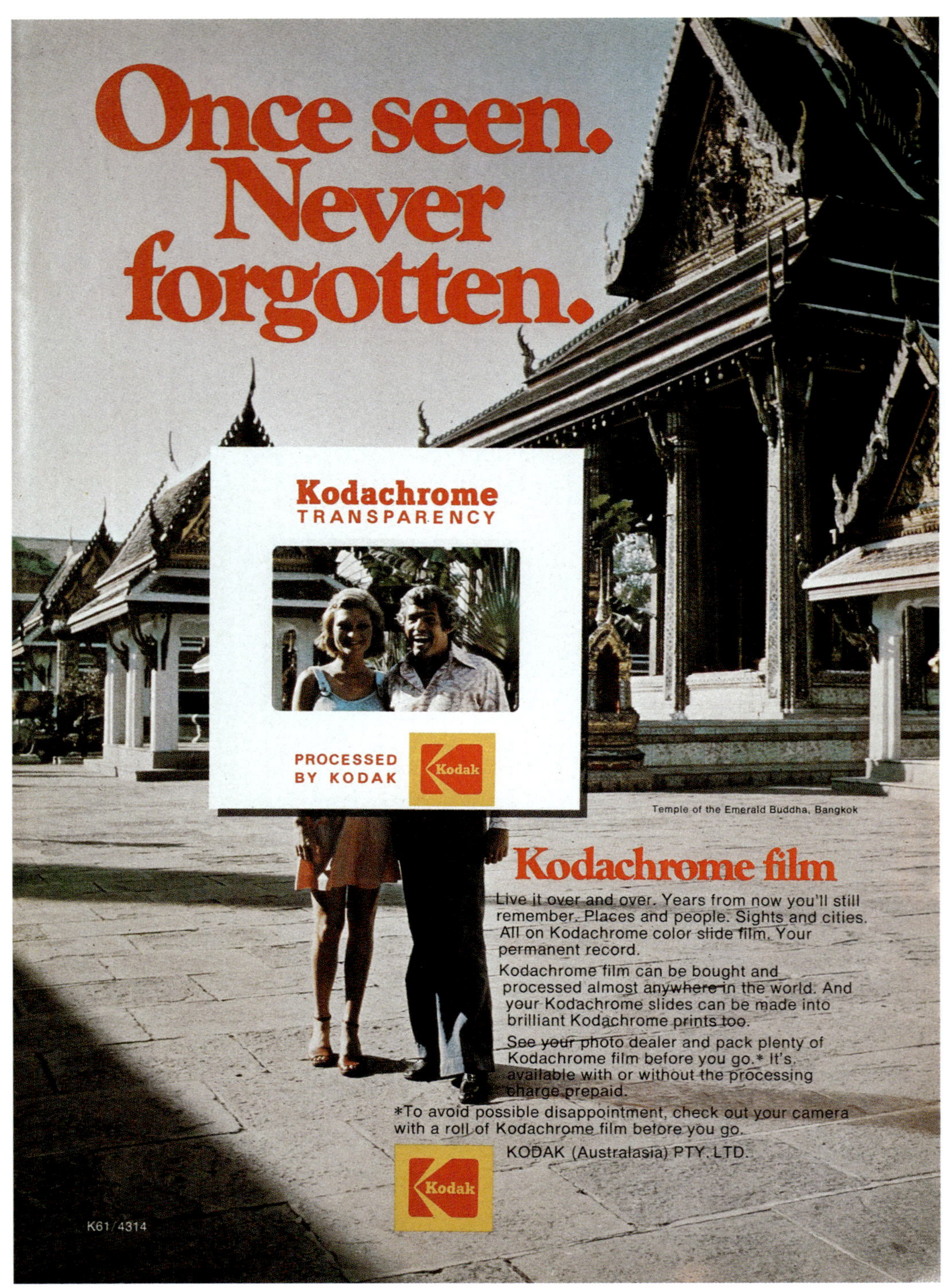

1973, Australia ▲   1972, Australia ▶

Your own pictures are the best travel souvenirs

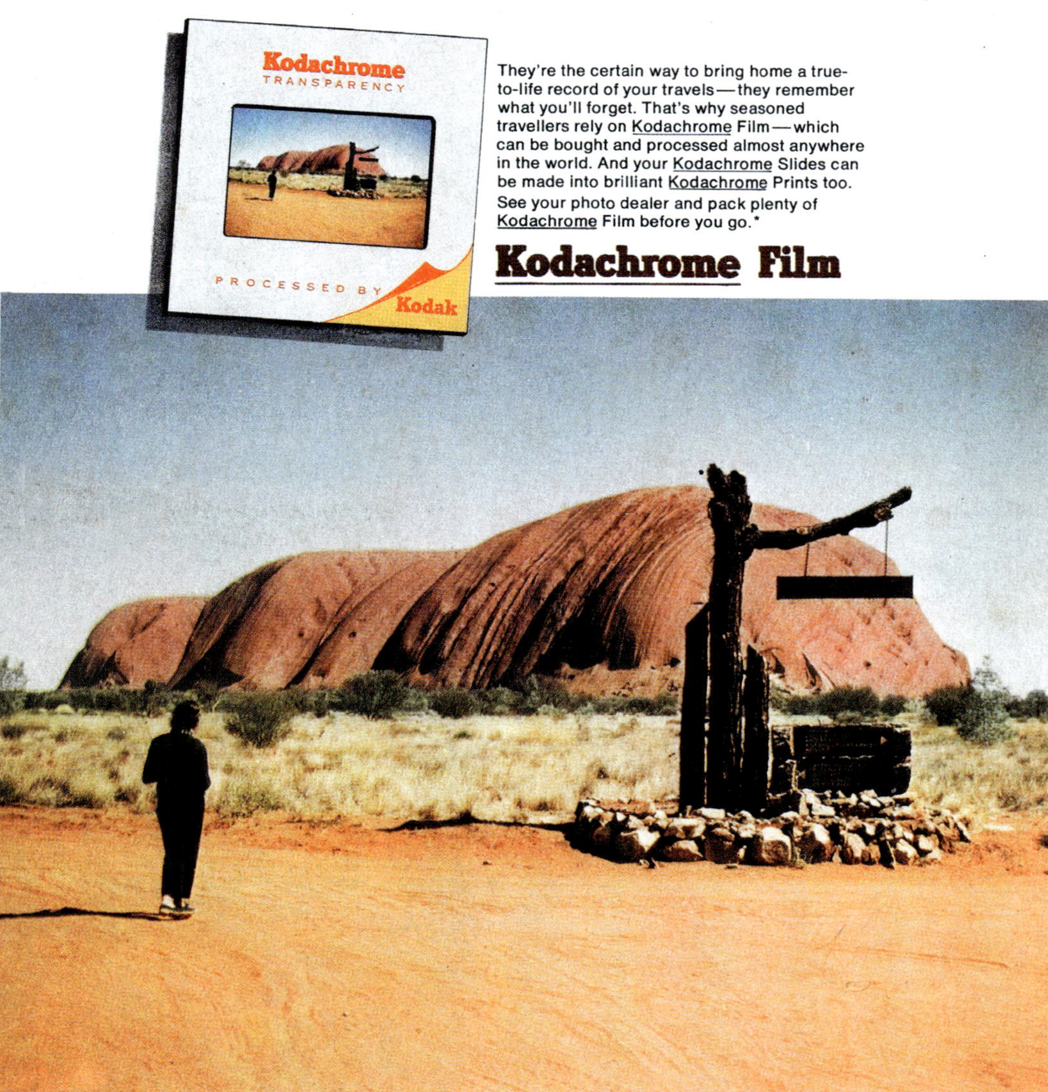

They're the certain way to bring home a true-to-life record of your travels — they remember what you'll forget. That's why seasoned travellers rely on Kodachrome Film — which can be bought and processed almost anywhere in the world. And your Kodachrome Slides can be made into brilliant Kodachrome Prints too. See your photo dealer and pack plenty of Kodachrome Film before you go.*

Kodachrome Film

Ayers Rock — Central Australia

*Before you go, take a few test pictures to check that your camera is operating correctly.

KODAK (Australasia) PTY. LTD.

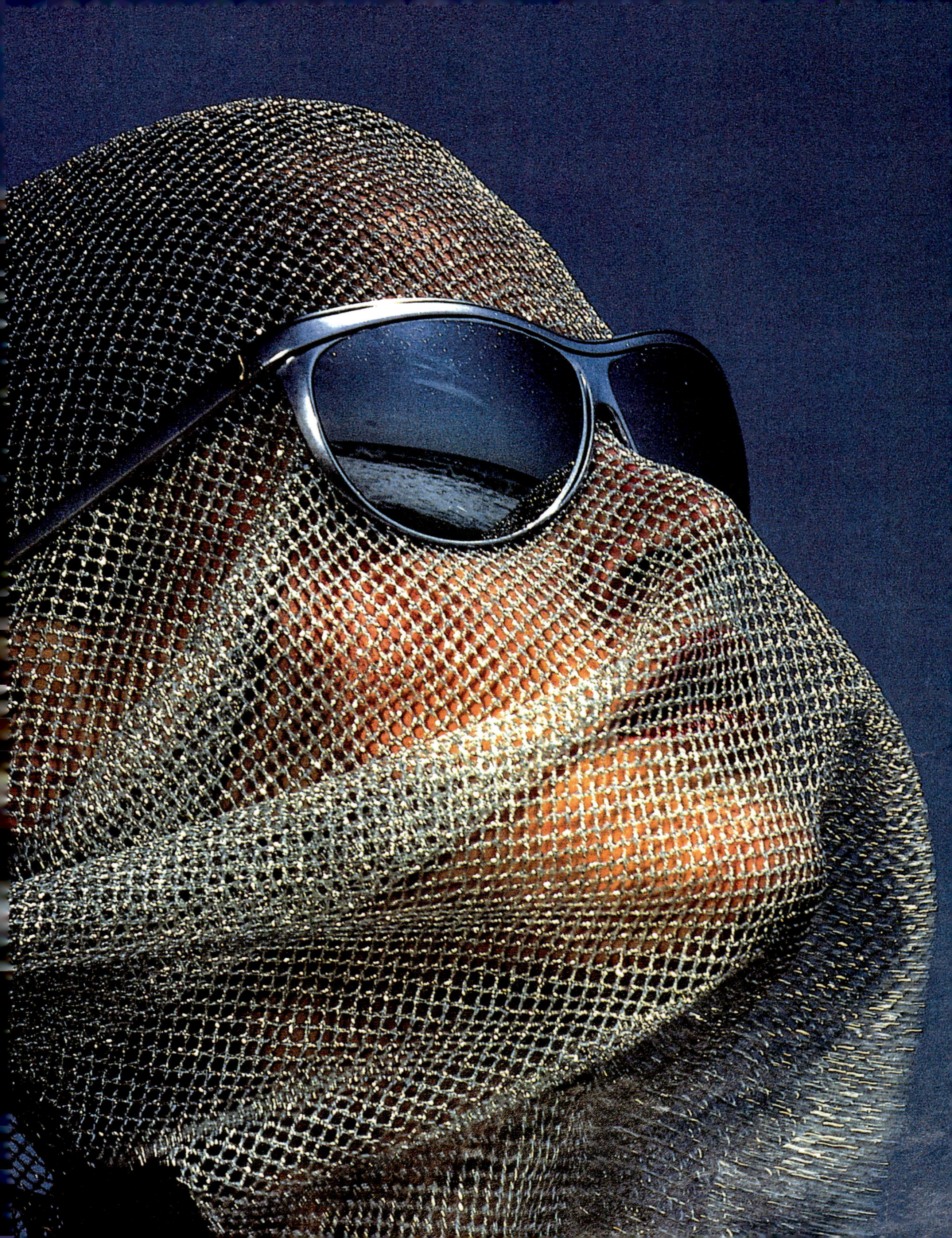

Creative Peak
1980-1991

In the early to mid-1980s, Kodak's advertising finally came into its own in terms of creative style. Rather than spill ink producing print ads that did little more than serve as visual pablum to sell cameras and projectors, Kodak's final wave of print advertising focused the consumer's mind's eye on bold visuals that fused Kodachrome's saturated color with thought-provoking imagery. The ads were bold and well received. Unfortunately, the analog film world would soon succumb to the rapid advances of digital imaging.

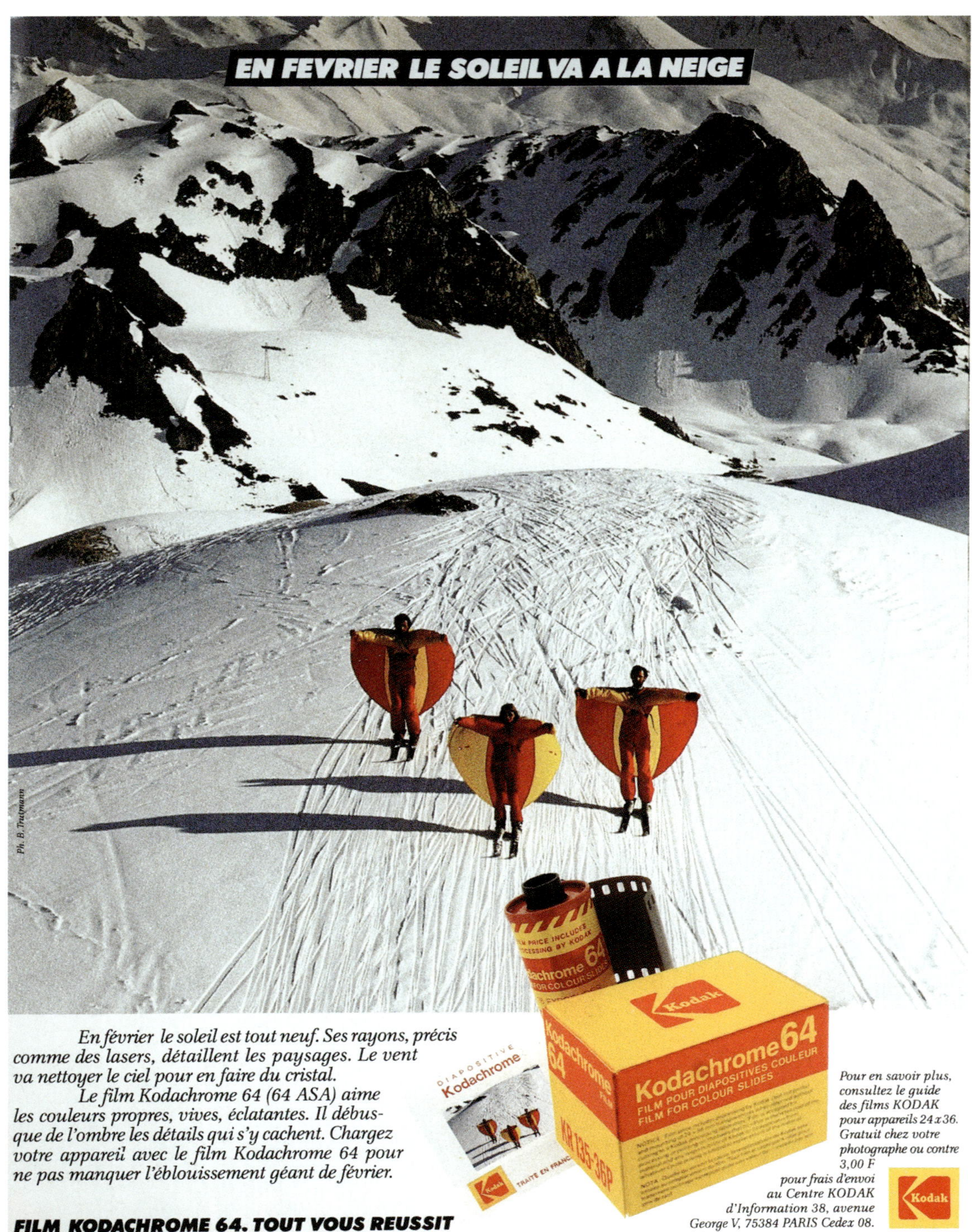

1982, France ▲   1983, USA ▶   1954, France ▶▶

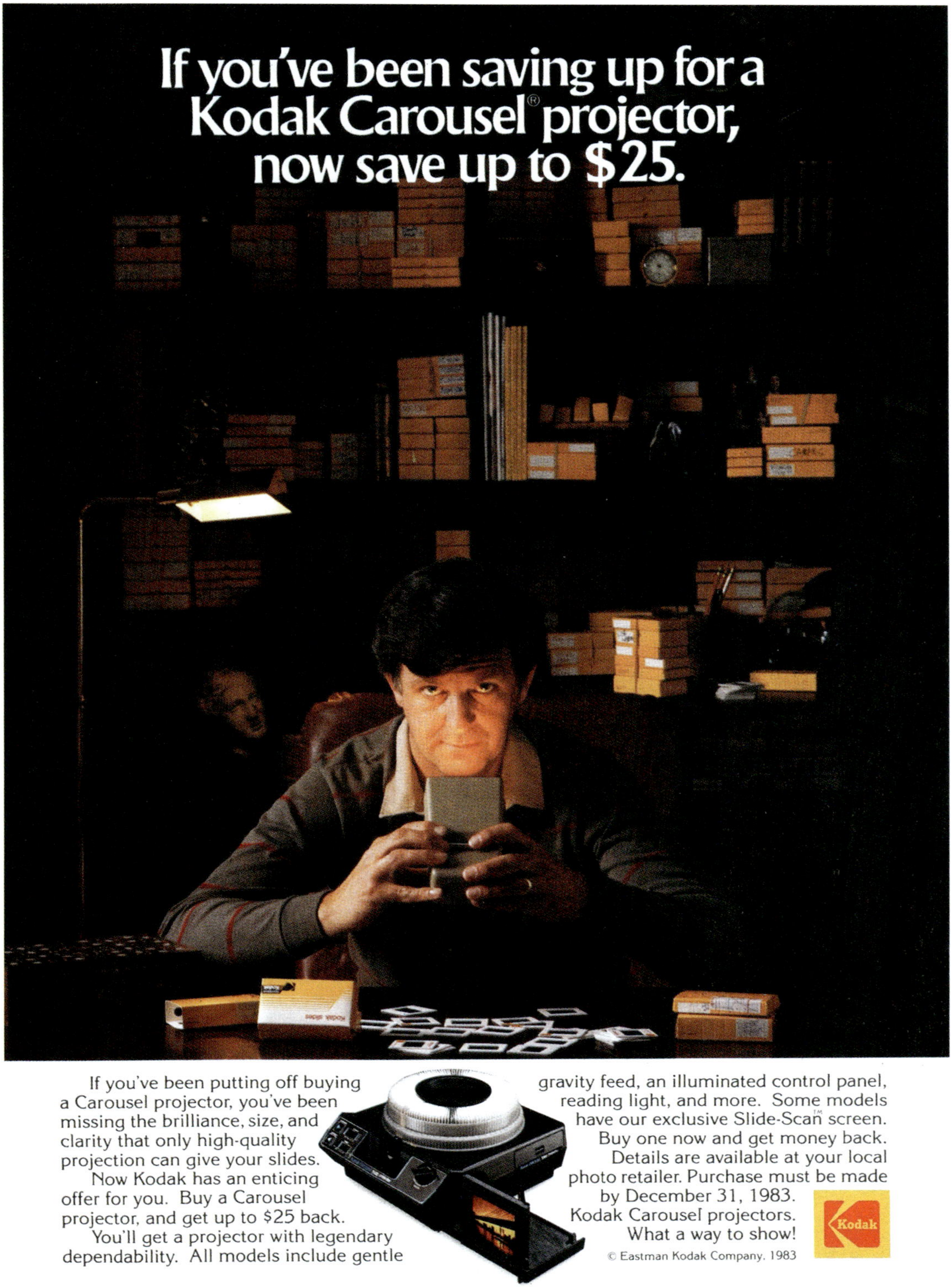

If you've been saving up for a Kodak Carousel® projector, now save up to $25.

If you've been putting off buying a Carousel projector, you've been missing the brilliance, size, and clarity that only high-quality projection can give your slides.

Now Kodak has an enticing offer for you. Buy a Carousel projector, and get up to $25 back.

You'll get a projector with legendary dependability. All models include gentle gravity feed, an illuminated control panel, reading light, and more. Some models have our exclusive Slide-Scan™ screen.

Buy one now and get money back. Details are available at your local photo retailer. Purchase must be made by December 31, 1983. Kodak Carousel projectors. What a way to show!

© Eastman Kodak Company. 1983

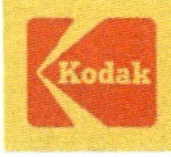

IL Y A UNE CENTAINE

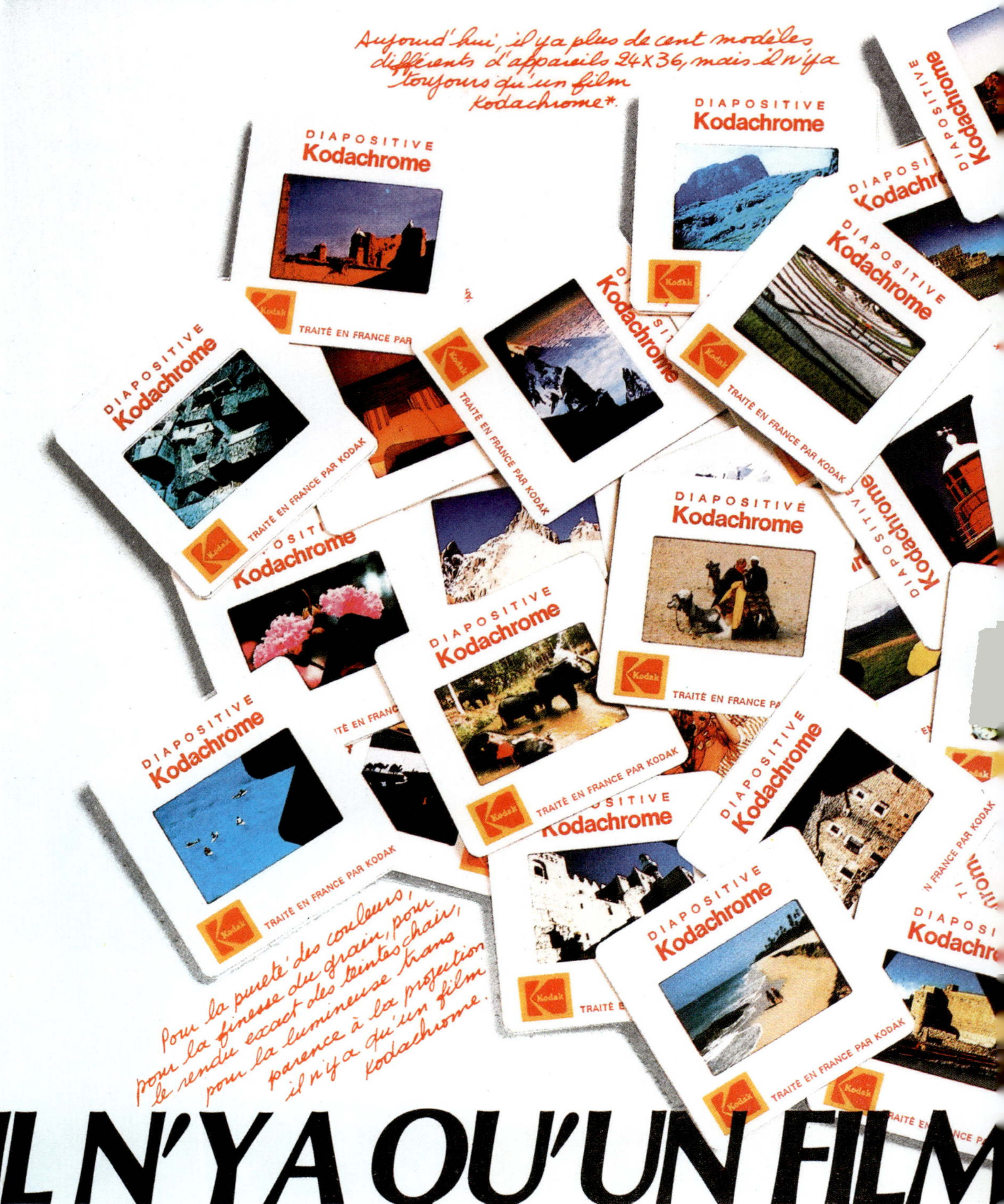

IL N'Y A QU'UN FILM

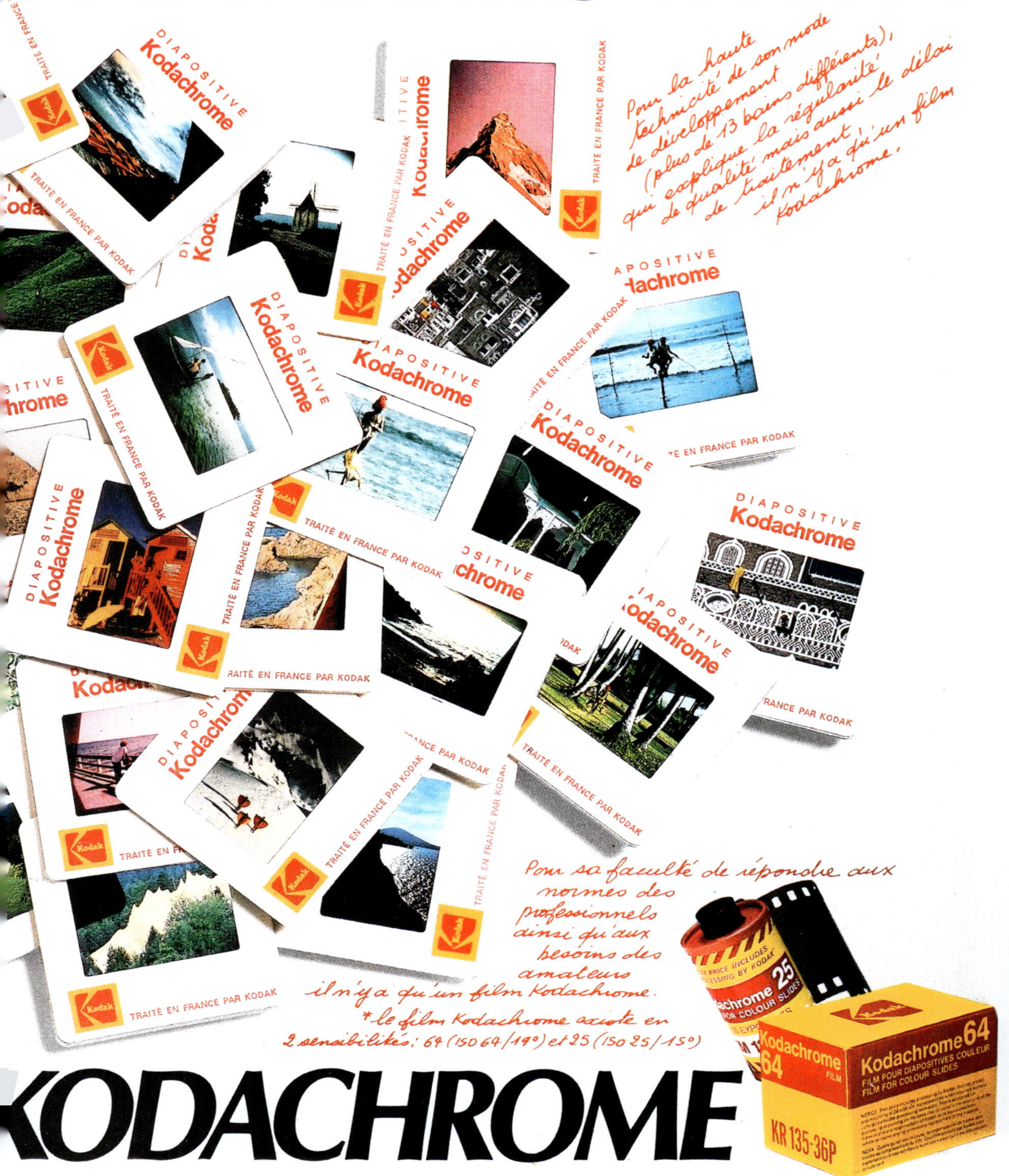

D'APPAREILS 24 X 36

Pour la haute
technicité de son mode
de développement
(plus de 13 bains différents),
qui explique la régularité
de qualité; mais aussi le délai
de traitement; il n'y a qu'un film
Kodachrome.

Pour sa faculté de répondre aux
normes des
professionnels
ainsi qu'aux
besoins des
amateurs
il n'y a qu'un film Kodachrome.
* le film Kodachrome existe en
2 sensibilités: 64 (ISO 64/19°) et 25 (ISO 25/15°)

KODACHROME

DIAPOSITIVE Kodachrome
TRAITÉ EN FRANCE PAR KODAK

Kodachrome64
KR 135-36P
FILM POUR DIAPOSITIVES COULEUR
FILM FOR COLOUR SLIDES

**In 1986, Kodak ran a series of print advertise-
ments featuring tight, dramatically lit portraits** of
some of the premier editorial and advertising photographers of the day. The portraits were taken by
the equally renowned photographer Michael O'Neill.

Included among these photographers were Ernst
Haas, Eric Meola, Pete Turner, Art Kane, Jay Maisel,
and others. In each advertisement, the photographer
expressed what he liked about Kodachrome, which
in the case of Ernst Haas, included the "special smell"
that emanated from each box of slides when they
returned from the lab. He "loved" it.

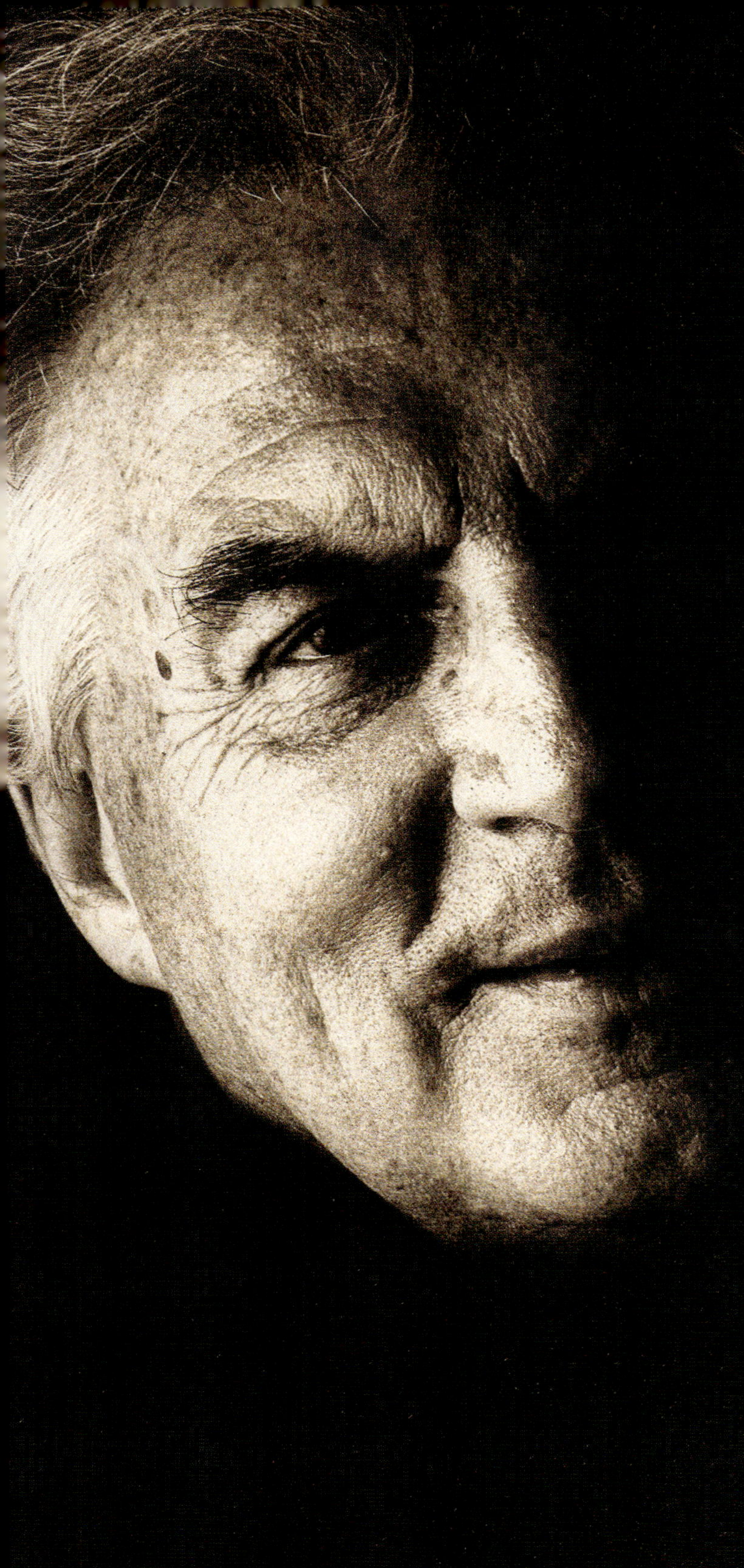

ERNST HAAS' 37-YEAR LOVE AFFAIR.

" I love Kodachrome film. When I started using it in 1949, it was a dream come true. Postwar Europe was gray and colorless, and I longed for color. Imagine how happy I was to find a color film that fit into my 35 mm camera and let me express what I wanted to.

" The original Kodachrome film had incredibly rich blacks and reds, and everything looked like a stained-glass window. Then Kodachrome II film came along, and it became a very beautiful film. Now Kodachrome 25 and 64 films are absolutely perfect.

" I still get a thrill when I return from a journey and get my Kodachrome film back from the lab. The slides have a special smell, and I love that smell. "

For more than 50 years, Kodachrome film has helped launch legends like Ernst Haas. Launch yours with Kodachrome professional film, distributed at tight tolerances for color balance and speed to meet your most critical photographic needs.

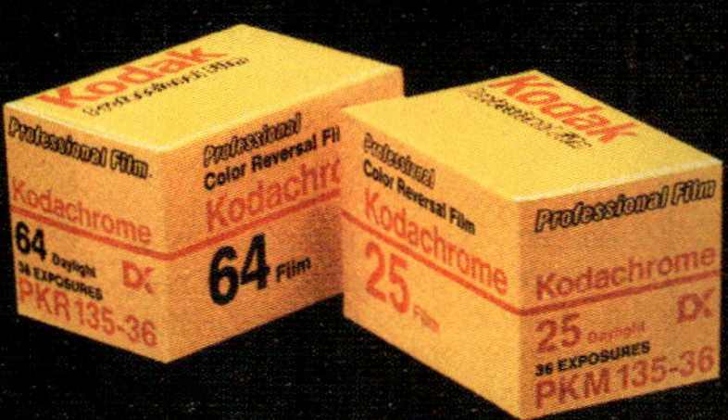

Kodak film—for a professional image.

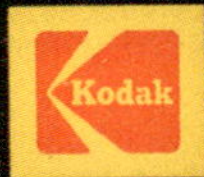

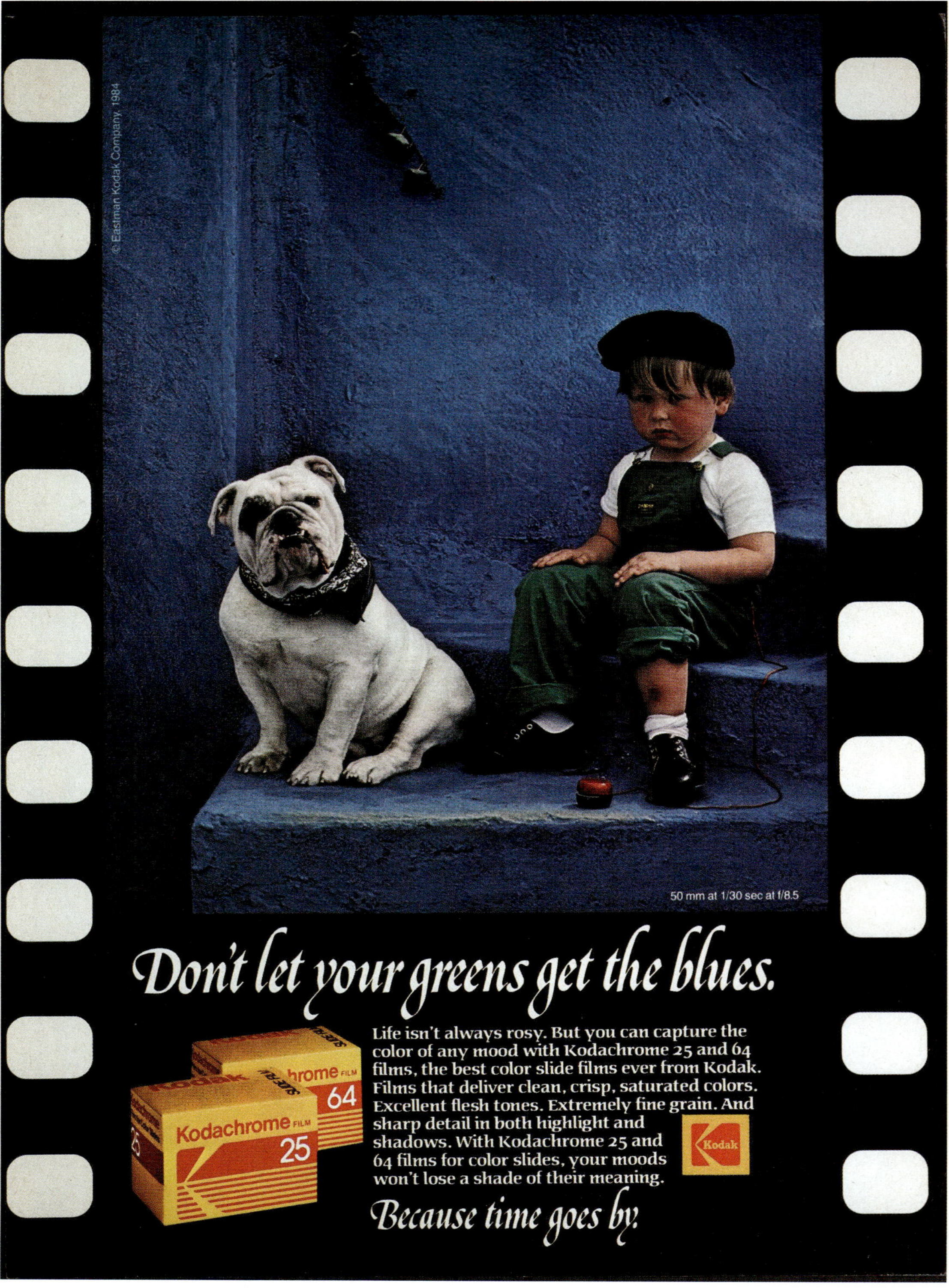

© Eastman Kodak Company, 1984
50 mm at 1/30 sec at f/8.5
Don't let your greens get the blues.
Kodachrome FILM 64
Kodachrome FILM 25
Kodak
Life isn't always rosy. But you can capture the color of any mood with Kodachrome 25 and 64 films, the best color slide films ever from Kodak. Films that deliver clean, crisp, saturated colors. Excellent flesh tones. Extremely fine grain. And sharp detail in both highlight and shadows. With Kodachrome 25 and 64 films for color slides, your moods won't lose a shade of their meaning.
Because time goes by.

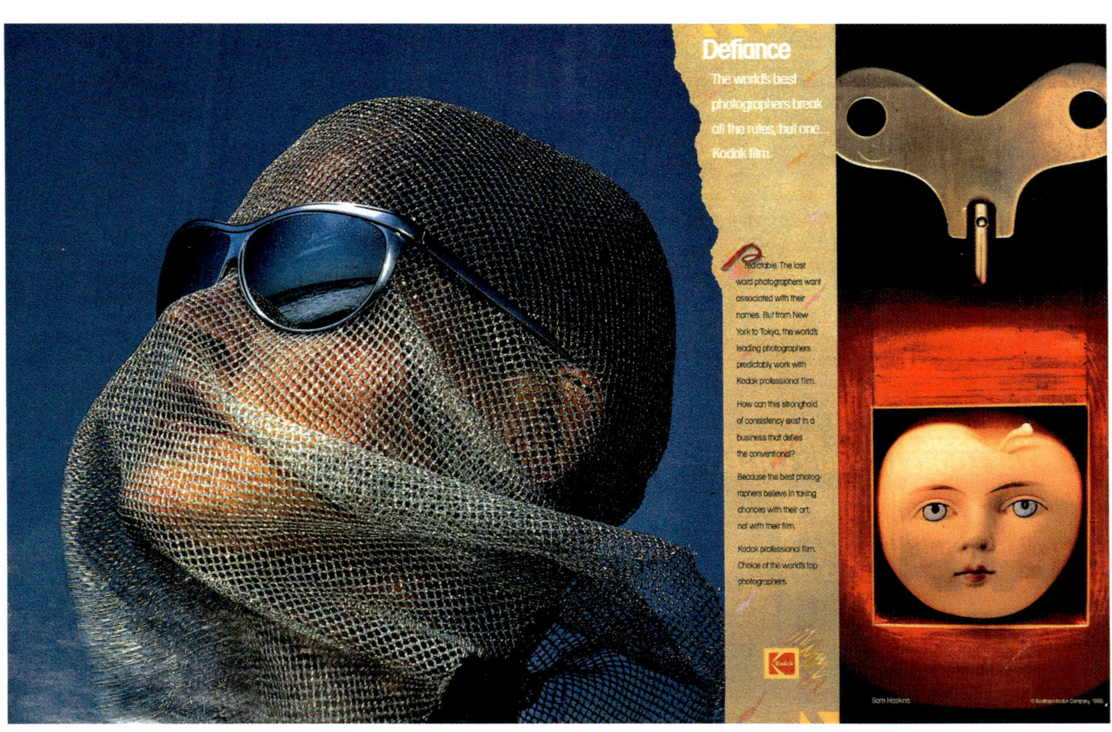

1986, USA   1986, USA

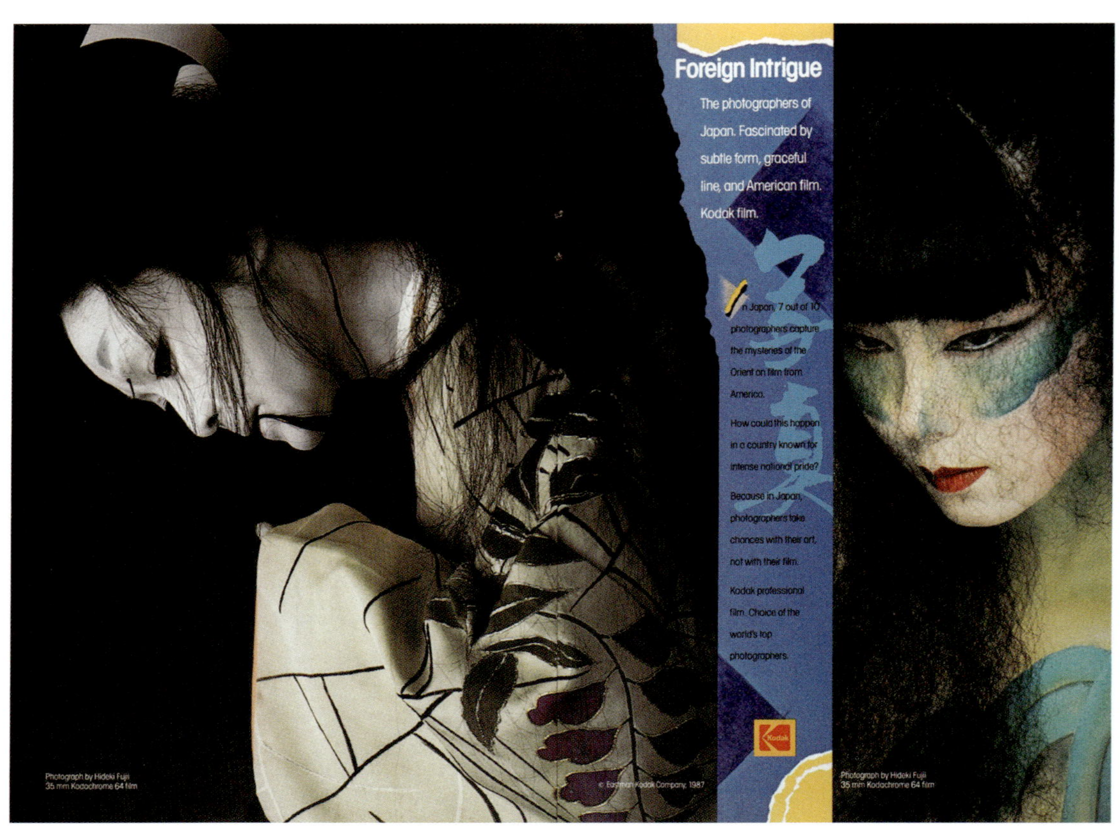

1987, USA   ▲   1987, USA   ▶

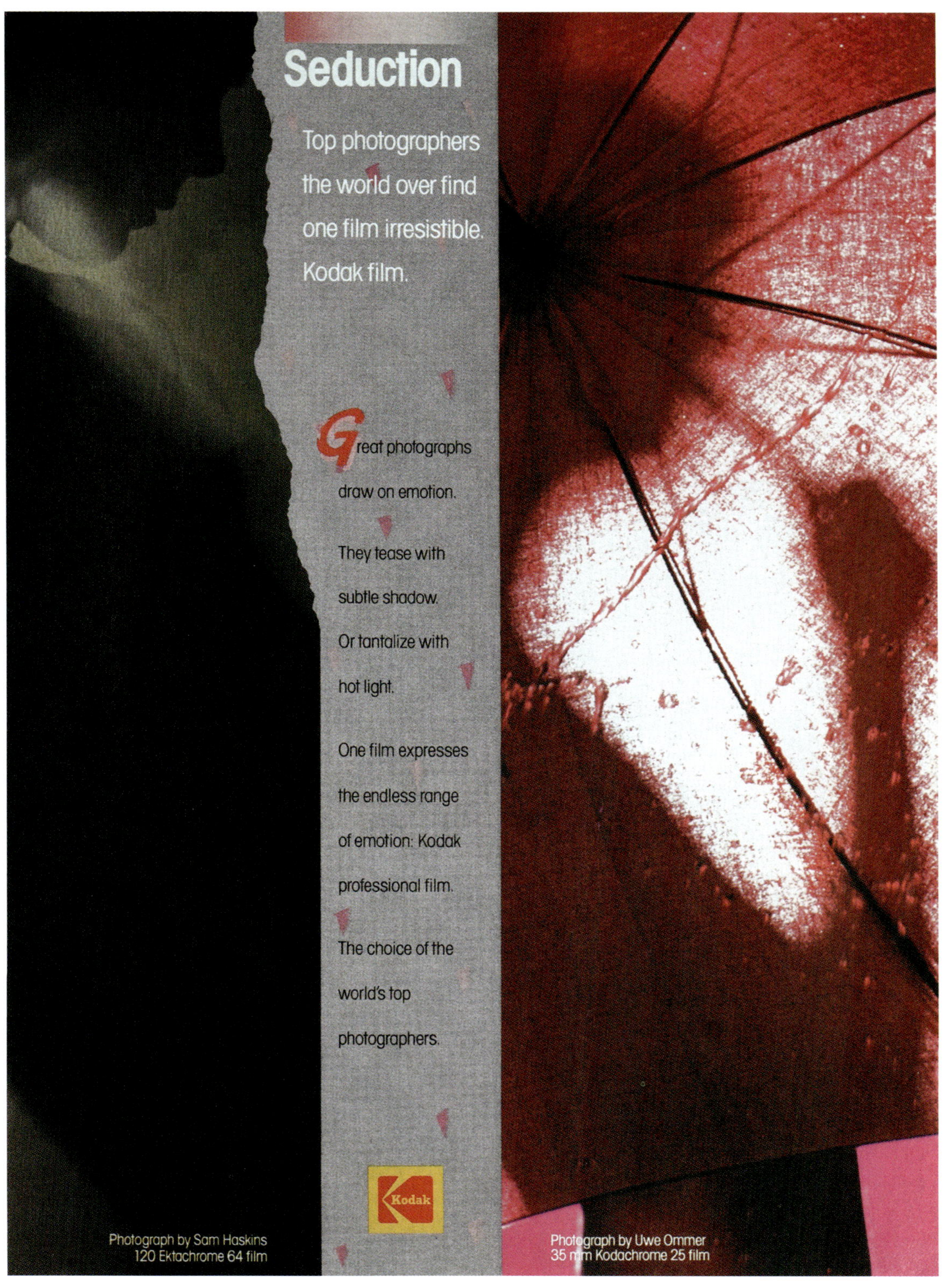

Seduction

Top photographers the world over find one film irresistible. Kodak film.

Great photographs draw on emotion.

They tease with subtle shadow. Or tantalize with hot light.

One film expresses the endless range of emotion: Kodak professional film.

The choice of the world's top photographers.

Kodak

Photograph by Sam Haskins
120 Ektachrome 64 film

Photograph by Uwe Ommer
35 mm Kodachrome 25 film

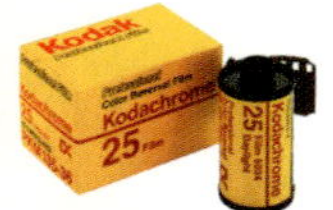

KODACHROME **25**

PROFESSIONAL FILM, 135-SIZE

This low-speed color reversal film (ISO 25) is balanced for daylight and features extremely fine grain, extremely high sharpness, high resolving power, and very saturated colors, particularly reds.

It's an excellent choice for commercial fashion, beauty, landscape, and product photography as well as for fine-art photography, industrial photography, and photojournalism requiring exceptionally sharp detail.

The recommended exposure range is 1/10,000 to 1/10 second. However, a 1-second exposure is possible by increasing the exposure 1/2 stop. The film is designed for exposure to daylight and electronic flash without filters, but you can also expose it using tungsten illumination (3200 K) with a No. 80A filter at ISO 6; or with photolamps (3400 K) and a No. 80B filter at ISO 8.

For outstanding results when taking photographs in open shade, aerial photographs, and photographs of distant views or sunlit snow scenes, use KODAK WRATTEN Gelatin Filter No. 1A (skylight filter) with no increase in exposure. A KODAK Light Balancing Filter No. 81B and a 1/3-stop exposure increase will help produce excellent results if your transparencies exposed by electronic flash are consistently too blue.

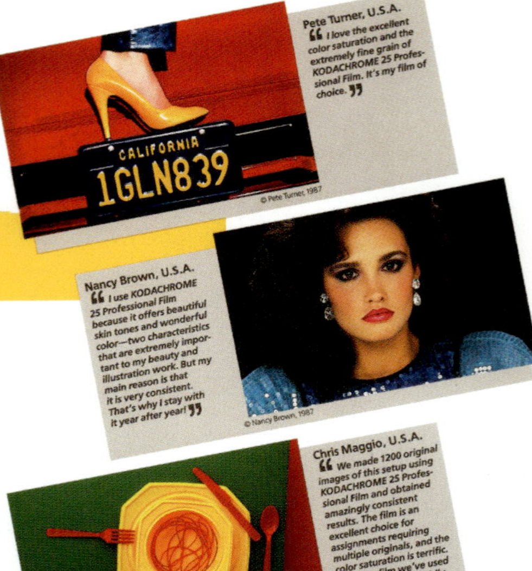

Pete Turner, U.S.A.
" I love the excellent color saturation and the extremely fine grain of KODACHROME 25 Professional Film. It's my film of choice. **"**

Nancy Brown, U.S.A.
" I use KODACHROME 25 Professional Film because it offers beautiful skin tones and wonderful color—two characteristics that are extremely important to my beauty and illustration work. But my main reason is that it is very consistent. That's why I stay with it year after year! **"**

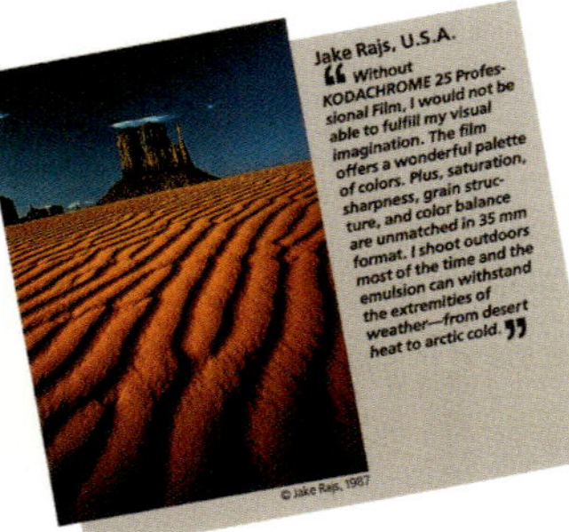

Jake Rajs, U.S.A.
" Without KODACHROME 25 Professional Film, I would not be able to fulfill my visual imagination. The film offers a wonderful palette of colors. Plus, saturation, sharpness, grain structure, and color balance are unmatched in 35 mm format. I shoot outdoors most of the time and the emulsion can withstand the extremities of weather—from desert heat to arctic cold. **"**

Chris Maggio, U.S.A.
" We made 1200 original images of this setup using KODACHROME 25 Professional Film and obtained amazingly consistent results. The film is an excellent choice for assignments requiring multiple originals, and the color saturation is terrific. No other film we've used can reproduce as vividly and accurately as KODACHROME Professional Film. **"**

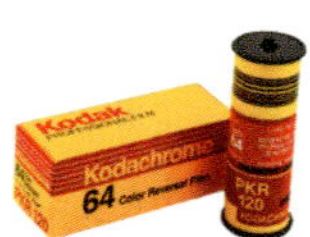

KODACHROME **64**

PROFESSIONAL FILM, 120-SIZE

This incredible medium-format, color reversal film (ISO 64) is daylight-balanced and features extremely fine grain, extremely high sharpness, high resolving power, and very saturated colors!

It's an excellent choice for commercial catalog, fashion, beauty, travel, and product photography as well as for outdoor-illustrative, fine-art, industrial photography, and photojournalism.

The recommended exposure range is 1/10,000 to 1/10 second. However, for a 1-second exposure increase the exposure 1 full stop and use a KODAK Color Compensating Filter CC 10R. The film is designed for exposure to daylight and electronic flash without filters. But you can also expose it using tungsten illumination (3200 K) with a No. 80A filter at ISO 16; or with photolamps (3400 K) and a No. 80B filter at ISO 20.

Aerial photographs, photographs taken in open shade, and photographs of distant views or sunlit snow scenes will yield best results when you use a KODAK WRATTEN Gelatin Filter No. 1A (skylight filter) which requires no increase in exposure. If your transparencies exposed by electronic flash are consistently too blue, you can get outstanding results by simply increasing exposure by 1/3 stop and attaching a KODAK Light Balancing Filter No. 81B.

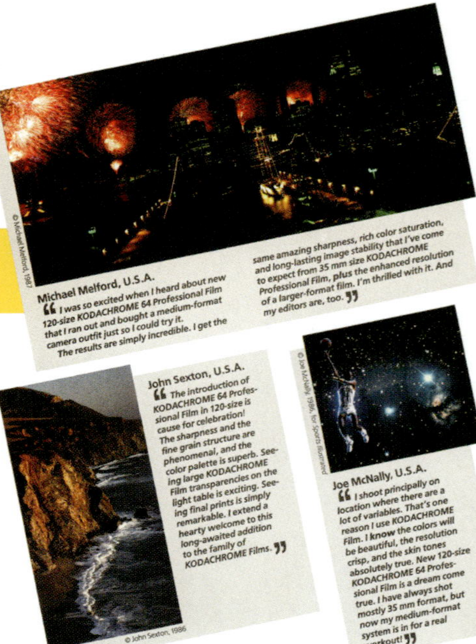

Michael Melford, U.S.A.
" I was so excited when I heard about new 120-size KODACHROME 64 Professional Film that I ran out and bought a medium-format camera outfit just so I could try it. The results are simply incredible. I get the same amazing sharpness, rich color saturation, and long-lasting image stability that I've come to expect from 35 mm size KODACHROME Professional Film, plus the enhanced resolution of a larger-format film. I'm thrilled with it. And my editors are, too. **"**

John Sexton, U.S.A.
" The introduction of KODACHROME 64 Professional Film in 120-size is cause for celebration! The sharpness and the fine grain structure are phenomenal, and the color palette is superb. Seeing large KODACHROME Film transparencies on the light table is exciting. Seeing final prints is simply remarkable. I extend a hearty welcome to this long-awaited addition to the family of KODACHROME Films. **"**

Joe McNally, U.S.A.
" I shoot principally on location where there are a lot of variables. That's one reason I use KODACHROME Film. I **know** the colors will be beautiful, the resolution crisp, and the skin tones absolutely true. New 120-size KODACHROME 64 Professional Film is a dream come true. I have always shot mostly 35 mm format, but now my medium-format system is in for a real workout! **"**

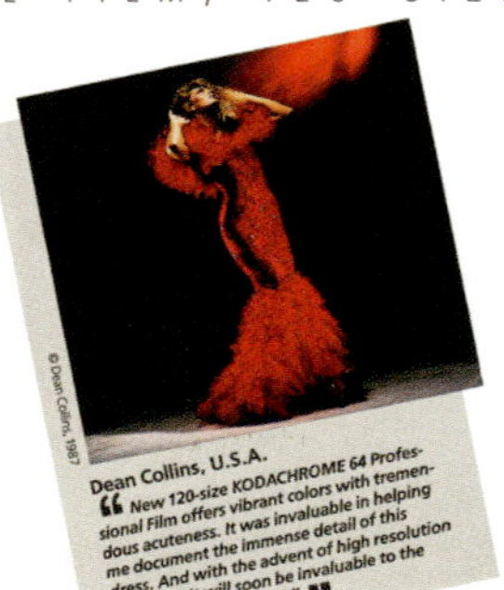

Dean Collins, U.S.A.
" New 120-size KODACHROME 64 Professional Film offers vibrant colors with tremendous acuteness. It was invaluable in helping me document the immense detail of this dress. And with the advent of high resolution scanning, it will soon be invaluable to the printing industry as well. **"**

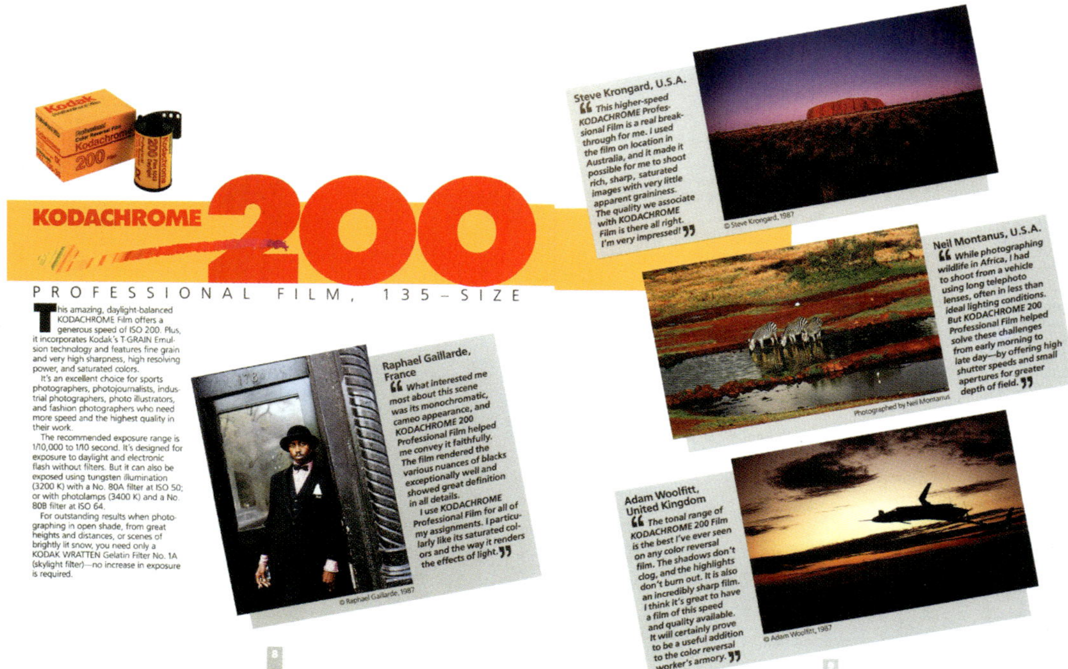

This amazing, daylight-balanced KODACHROME Film offers a generous speed of ISO 200. Plus, it incorporates Kodak's T-GRAIN Emulsion technology and features fine grain and very high sharpness, high resolving power, and saturated colors.

It's an excellent choice for sports photographers, photojournalists, industrial photographers, photo illustrators, and fashion photographers who need more speed and the highest quality in their work.

The recommended exposure range is 1/10,000 to 1/10 second. It's designed for exposure to daylight and electronic flash without filters. But it can also be exposed using tungsten illumination (3200 K) with a No. 80A filter at ISO 50; or with photolamps (3400 K) and a No. 80B filter at ISO 64.

For outstanding results when photographing in open shade, from great heights and distances, or scenes of brightly lit snow, you need only a KODAK WRATTEN Gelatin Filter No. 1A (skylight filter)—no increase in exposure is required.

Steve Krongard, U.S.A. " This higher-speed KODACHROME Professional Film is a real breakthrough for me. I used the film on location in Australia, and it made it possible for me to shoot rich, sharp, saturated images with very little apparent graininess. The quality we associate with KODACHROME Film is there all right. I'm very impressed! "

© Steve Krongard, 1987

Raphael Gaillarde, France " What interested me most about this scene was its monochromatic, cameo appearance, and KODACHROME 200 Professional Film helped me convey it faithfully. The film rendered the various nuances of blacks exceptionally well and showed great definition in all details. I use KODACHROME Professional Film for all of my assignments. I particularly like its saturated colors and the way it renders the effects of light. "

© Raphael Gaillarde, 1987

Neil Montanus, U.S.A. " While photographing wildlife in Africa, I had to shoot from a vehicle using long telephoto lenses, often in less than ideal lighting conditions. But KODACHROME 200 Professional Film helped solve these challenges from early morning to late day—by offering high shutter speeds and small apertures for greater depth of field. "

Photographed by Neil Montanus

Adam Woolfitt, United Kingdom " The tonal range of KODACHROME 200 Film is the best I've ever seen on any color reversal film. The shadows don't clog, and the highlights don't burn out. It is also an incredibly sharp film. I think it's great to have a film of this speed and quality available. It will certainly prove to be a useful addition to the color reversal worker's armory. "

© Adam Woolfitt, 1987

As the 1980s led into the 1990s, the decline of film and ascent of digital imaging was becoming increasingly obvious to all. Regardless, Kodak's advertising at this juncture was perhaps the most visually creative period of time for Kodachrome, with exciting imagery being captured in the street as well as in the studio.

During these final years, Kodak made a point of hiring and giving photo credits to some of the most creative photographers of the day. Included among these luminaries were Sam Haskins, Uwe Ommer, Nancy Brown, Chris Maggio, Jake Rajs, Pete Turner, Dean Collins, John Sexton, Michael Medford, Sarah Moon, and Joe McNally.

▲ 1987

◀ 1987

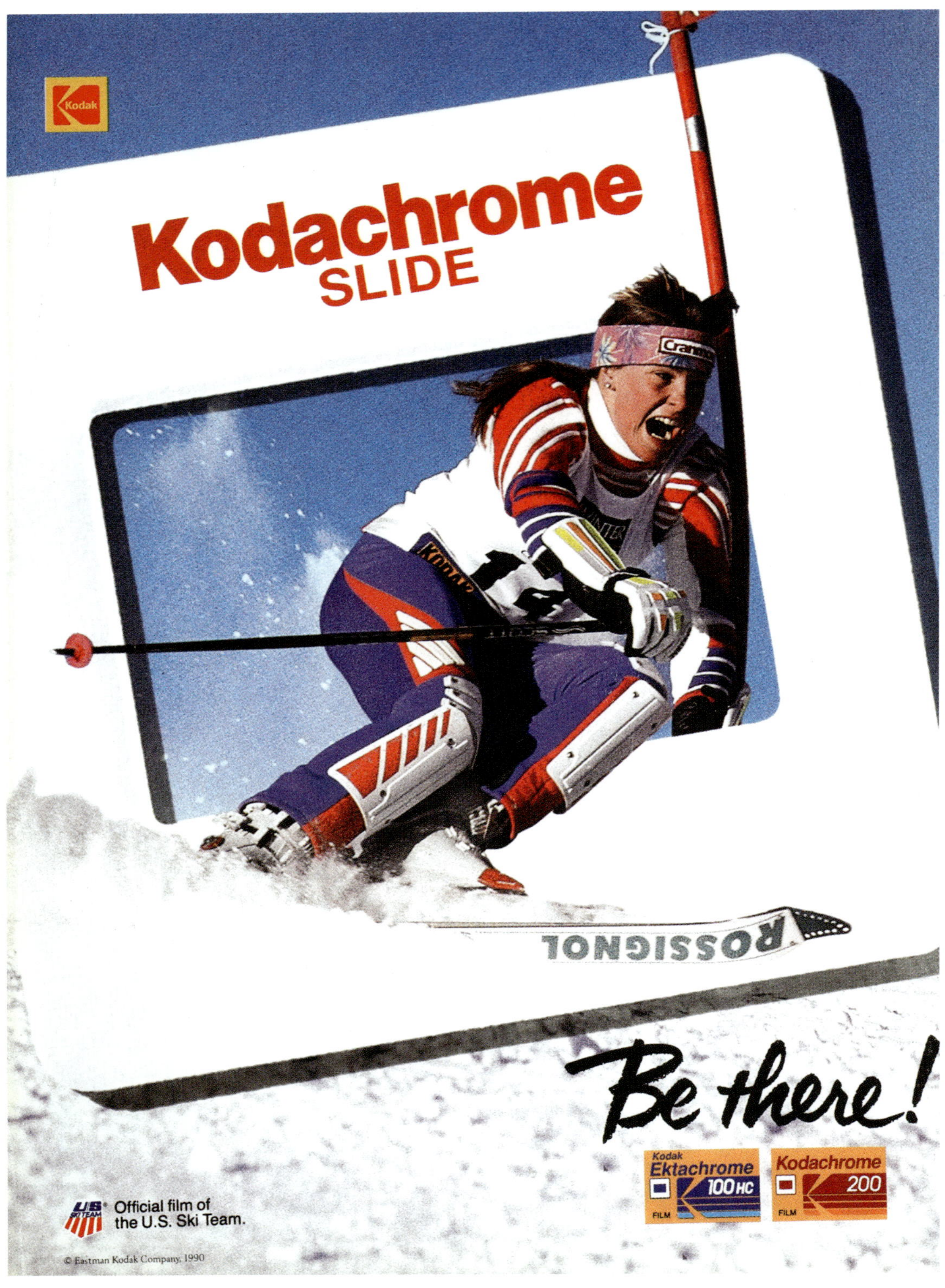
Kodak
Kodachrome
SLIDE
© Eastman Kodak Company, 1990
ROSSIGNOL
Be there!
Official film of
the U.S. Ski Team.
Kodak
Ektachrome
100 HC
FILM
Kodachrome
200
FILM
© Eastman Kodak Company, 1990

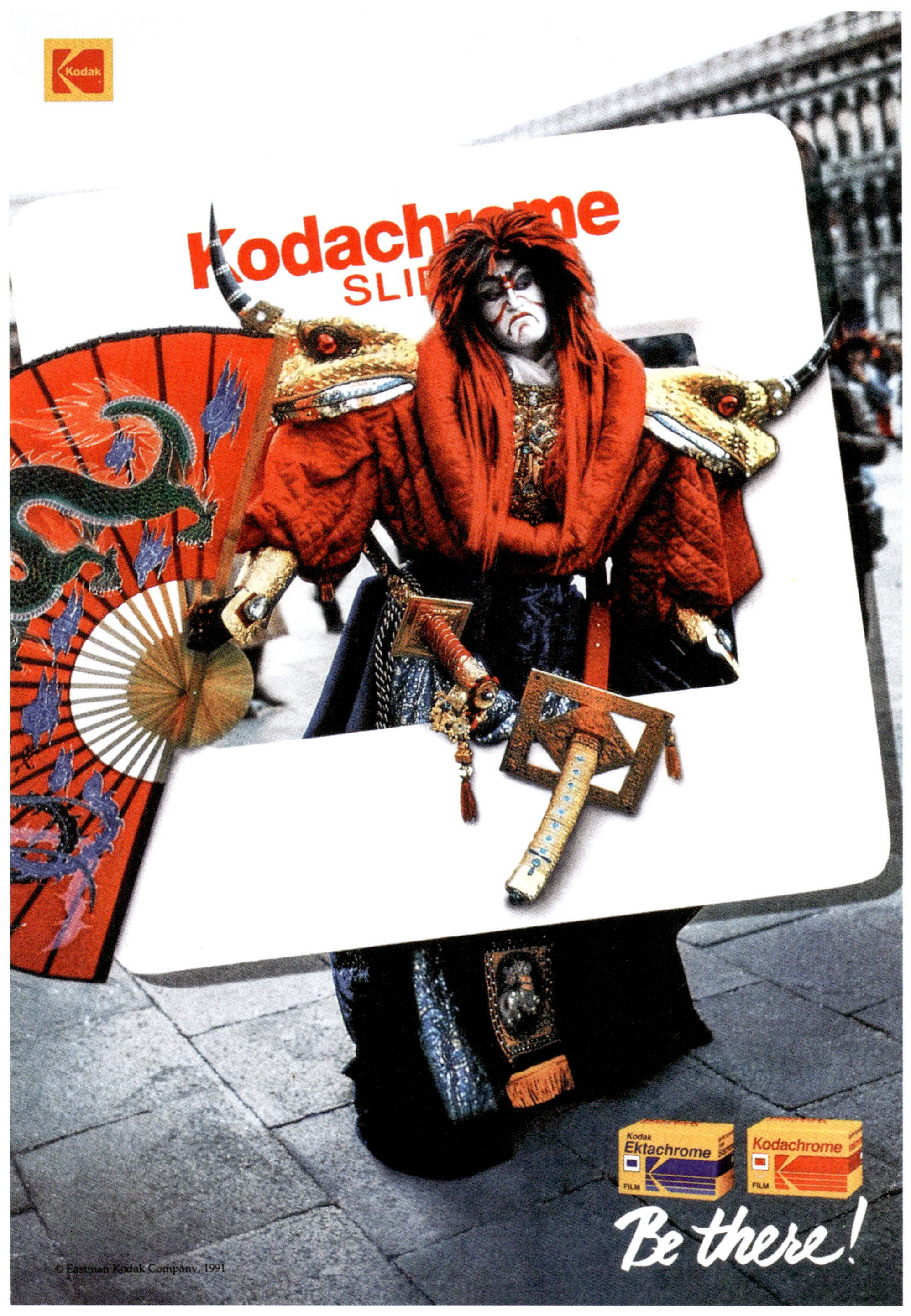

◀ 1990, USA   ⬟ 1991, USA

In 1985, Nikon produced a five-page magazine advertising supplement introducing its newest 35mm film cameras. Anybody want to guess what film the photographer is holding?

1985 Nikon Ad

It's only as good as the camera you put it in.

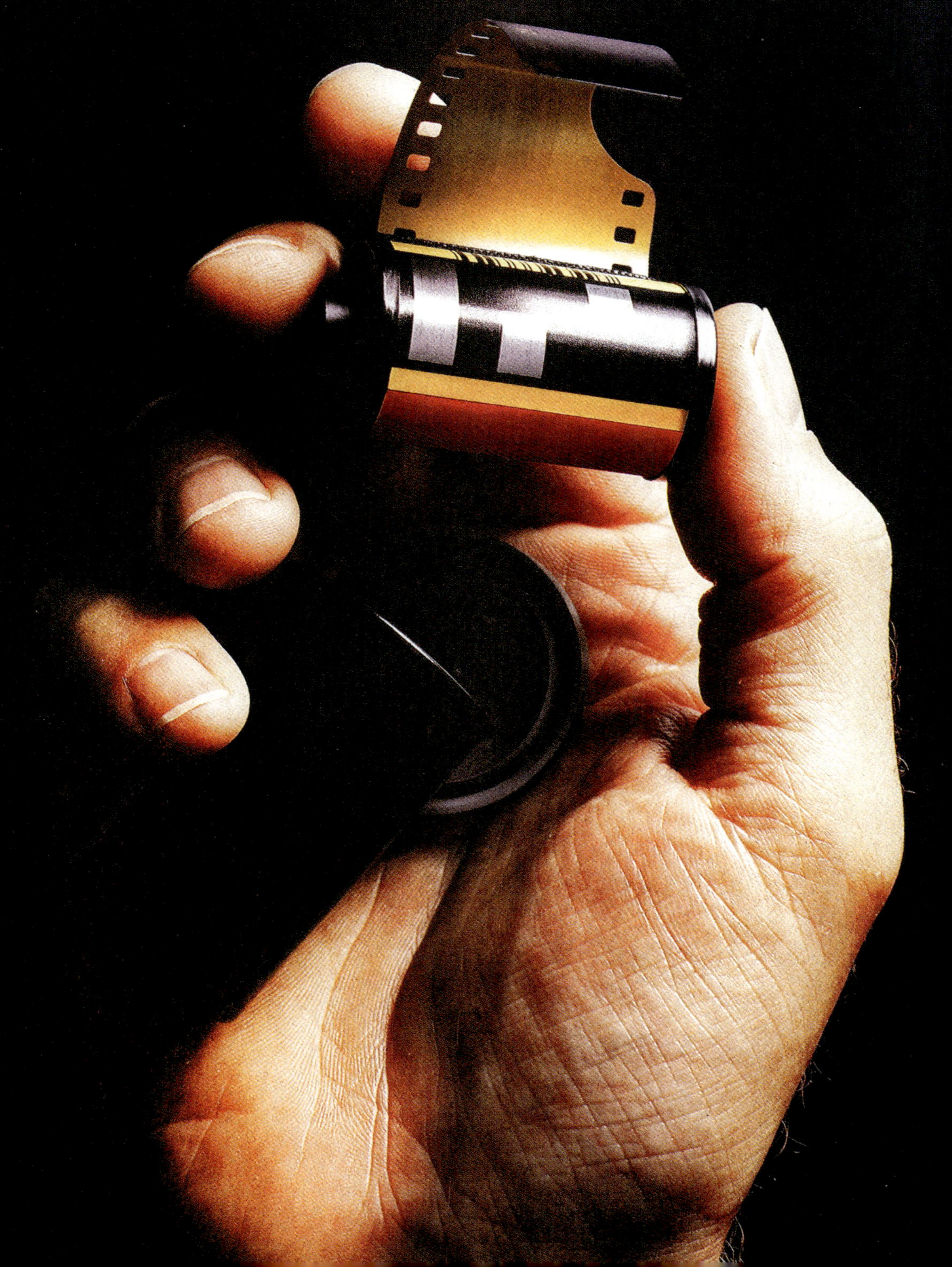

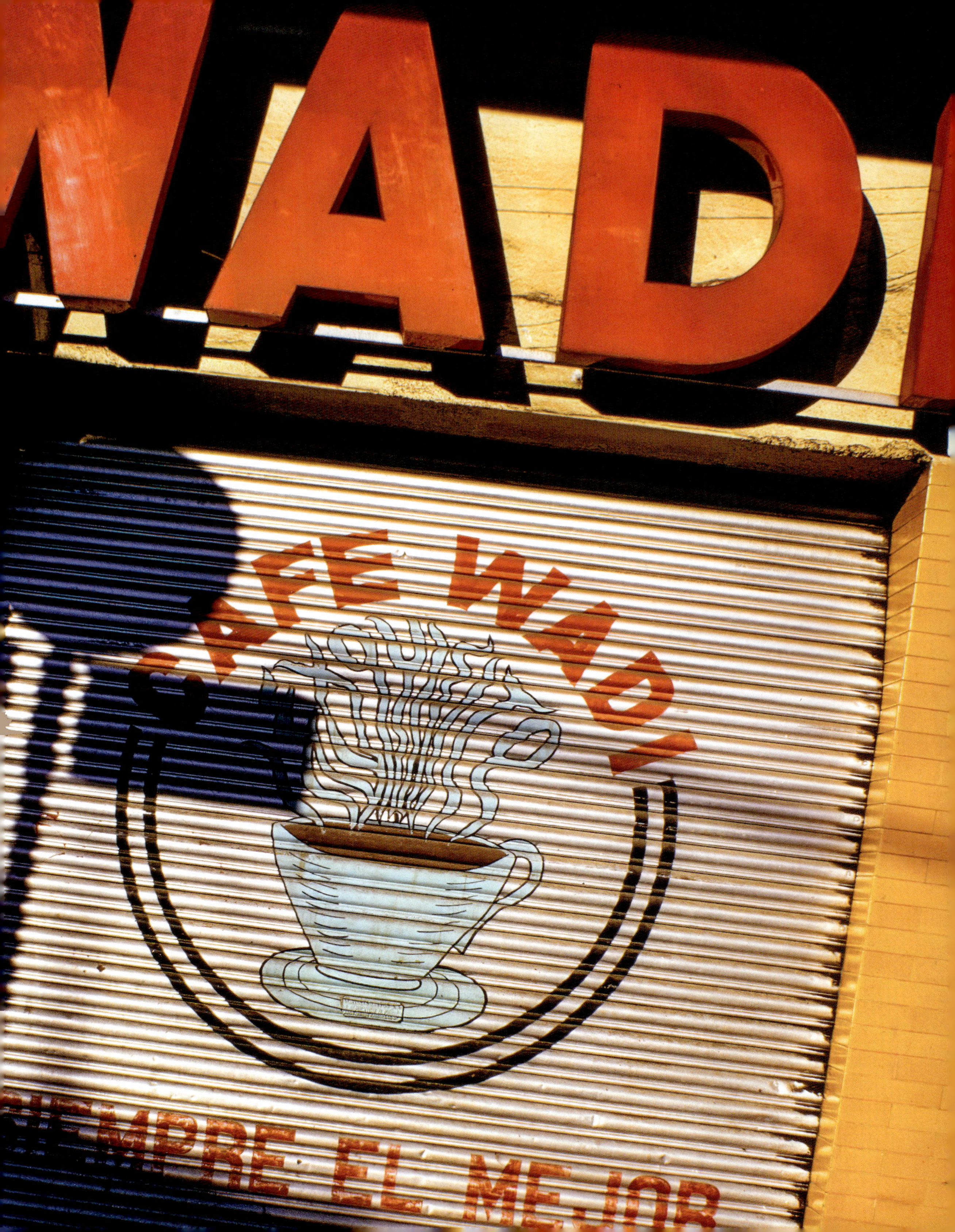
WAD
CAFE WADI
SIEMPRE EL MEJOR

SONGS & STATE PARKS

Kodachrome's Cultural Footprint

Kodachrome By...

Following its release in 1935, Kodachrome was quickly adopted as the film of choice by National Geographic Magazine photographers. Before long, photo credit lines in National Geographic began reading "Kodachrome by..." rather than the typical "Photograph by..." That's a brass ring to which few films can or will ever be able to lay claim.

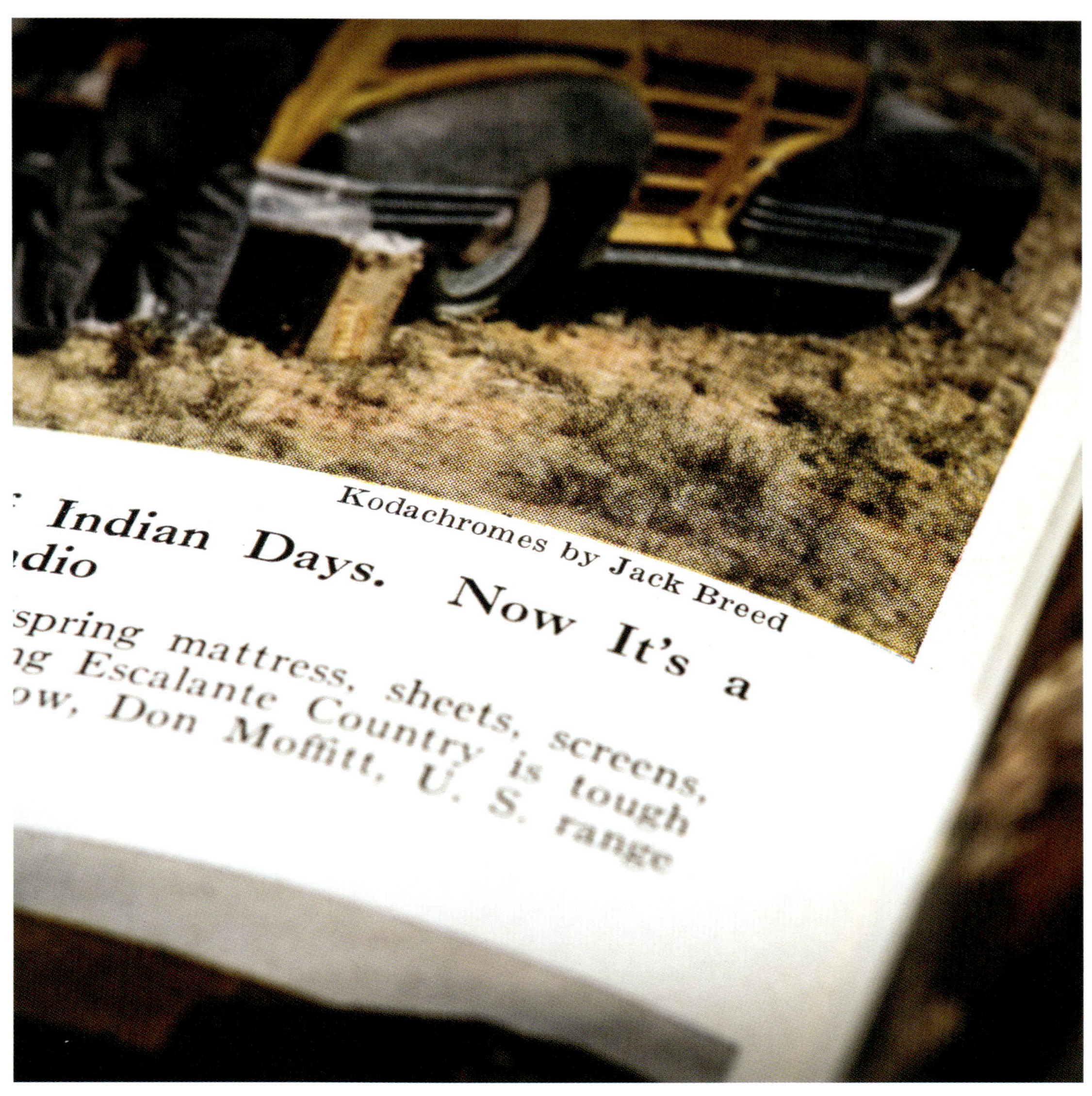

Kodachromes by Jack Breed
Indian Days. Now It's a
dio
spring mattress, sheets, screens,
Escalante Country is tough
w, Don Moffitt, U. S. range

Kodachrome: The Song

The song "Kodachrome" was the lead single from Paul Simon's third solo album titled, "There Goes Rhymin' Simon." The album was produced by Phil Ramone and released on May 19, 1973. To date, more than twenty-seven million copies of the album have been sold.

There's a nice touch of irony in that the film that brought us "nice bright colors, the greens of summers," and made "all the world a sunny day," was introduced in the midst of the Great Depression, a time and place usually viewed in grittier shades of gray.

Fun Facts about Paul Simon's "Kodachrome"

1. According to Paul Simon, the original title of "Kodachrome" was "Going Home." The opening lyrics of the song seem to concur.

2. Even though the song reached Number 2 in the US on Billboard's Hot 100 list, the song was never played on the airwaves in the UK due to a BBC ban on playing songs with branded products in the lyrics.

3. In a bid to prevent any legal action from Kodak against Paul Simon, Columbia Records, or anybody else involved in producing and/or promoting the song, a sticker reading "Kodachrome is a registered trademark of the Eastman Kodak Company" was placed on the cover of every record sold.

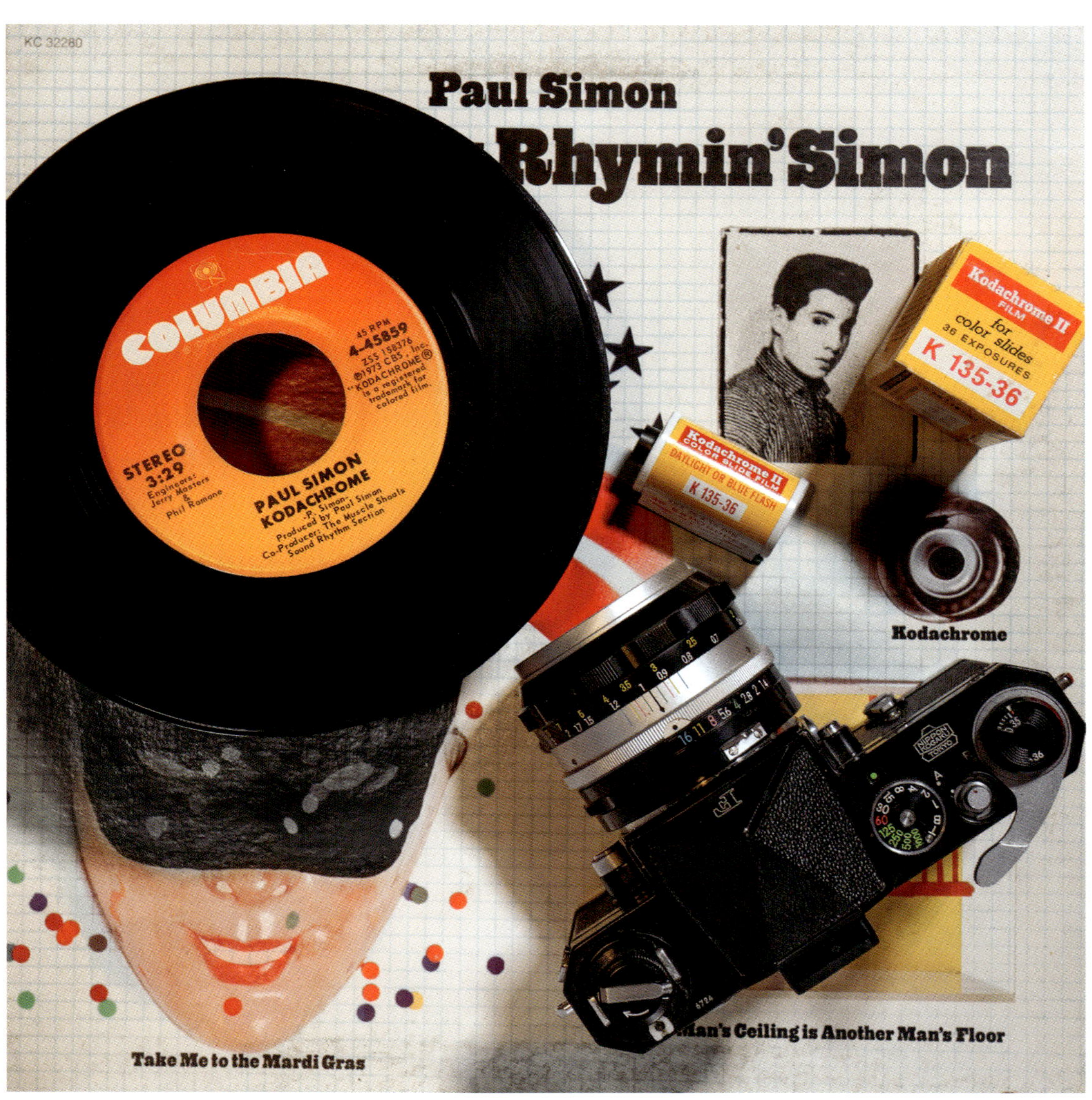

KC 32280
Paul Simon
Rhymin' Simon
COLUMBIA
45 RPM
4-45859
ZSS 158376
℗ 1973 CBS Inc.
"KODACHROME®
is a registered
trademark for
colored film.
STEREO
3:29
Engineers:
Jerry Masters
&
Phil Ramone
PAUL SIMON
KODACHROME
–P. Simon–
Produced by Paul Simon
Co-Producers: The Muscle Shoals
Sound Rhythm Section
Kodachrome II
FILM
for
color slides
36 EXPOSURES
K 135-36
Kodachrome II
COLOR SLIDE FILM
DAYLIGHT OR BLUE FLASH
K 135-36
NIPPON KOGAKU
TOKYO
Kodachrome
Take Me to the Mardi Gras
Man's Ceiling is Another Man's Floor

Kodachrome: The Movie

Produced for Netflix in 2018, "Kodachrome" is the story of the rekindling of the damaged relationship between a fictitious legendary photographer, Benjamin Asher Ryder (Ed Harris), and his estranged son Matt Ryder (Jason Sudeikis). Benjamin is terminally ill, and Matt has nothing but sour memories of their father/son relationship.

The story line follows the two as they drive cross-country to deliver Harris's last rolls of Kodachrome for development in what would be the very last processing run at Dwayne's Photo, in Parsons, Kansas (see page 147).

The story, which is intertwined with true events leading up to the final run of Kodachrome at Dwayne's Photo, has no shortage of humor and pain as the father and son attempt to mend their relationship, all while making sure they don't miss the last run of Kodachrome.

"Kodachrome," which was based on a New York Times article by A.G. Sulzberger, was written by Jonathan Tropper and directed by Mark Raso.

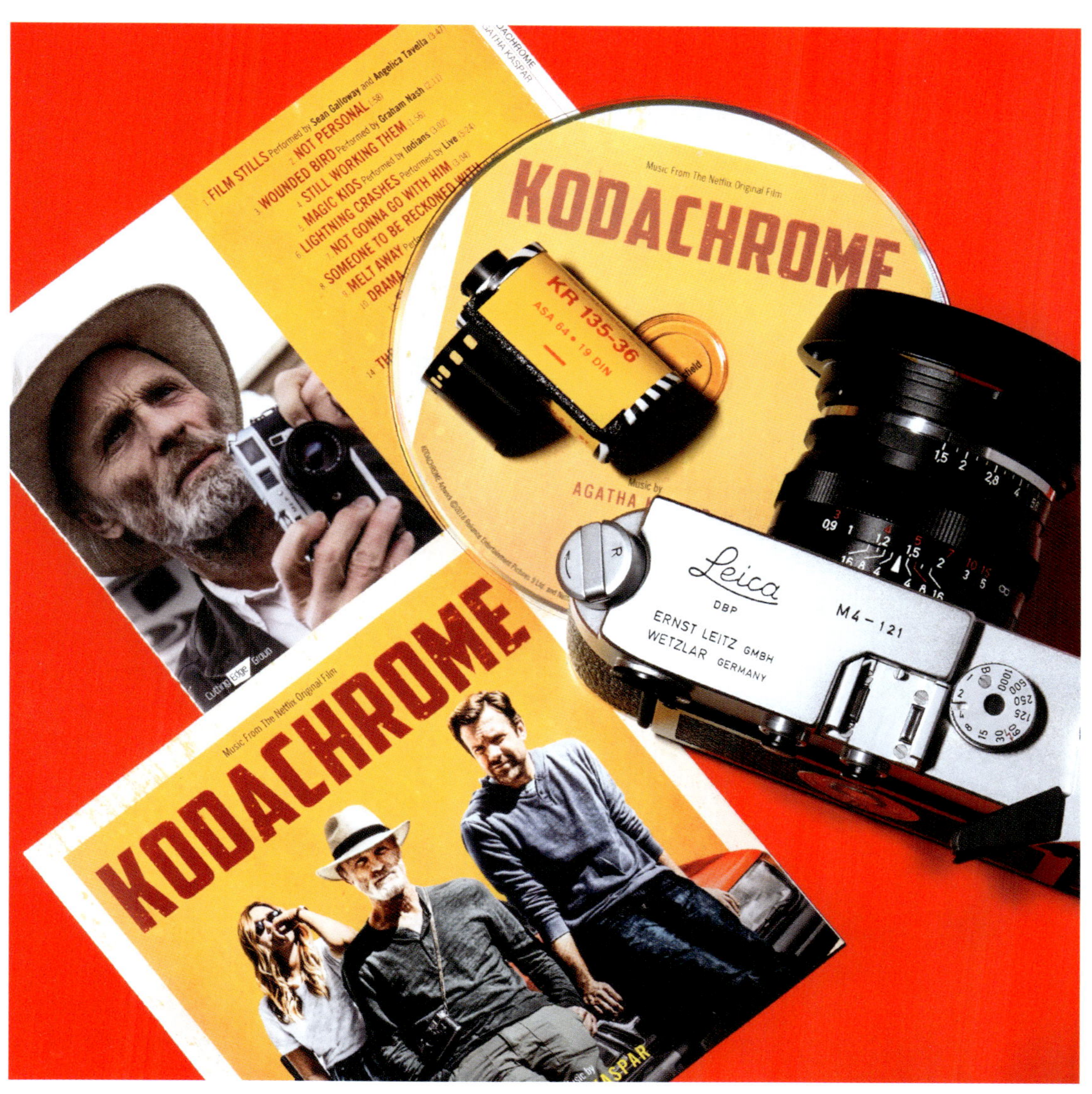

Music From The Netflix Original Film
KODACHROME
Music by
AGATHA
KR 135-36
ASA 64 • 19 DIN
1. FILM STILLS Performed by Sean Galloway and Angelica Tavella
2. NOT PERSONAL Performed by
3. WOUNDED BIRD Performed by Graham Nash
4. STILL WORKING THEM
5. MAGIC KIDS Performed by
6. LIGHTNING CRASHES Performed by Indians
7. NOT GONNA GO WITH HIM Performed by Live
8. SOMEONE TO BE RECKONED WITH
9. MELT AWAY
10. DRAMA
KODACHROME
Music From The Netflix Original Film
Leica
DBP
ERNST LEITZ GMBH
WETZLAR GERMANY
M4 - 121

Kodachrome: The State Park

The big attraction at Utah's Kodachrome Basin State Park are the sixty-seven colorful monolithic stone sedimentary pipes, or spires, which measure approximately 175 feet in height and are estimated to be about 180 million years old.

The National Geographic Society suggested naming the park after the film as a nod to the unique film-like color palette of the park's features, after photographing the area in 1948 for a feature article that was published in the September 1949 issue of National Geographic Magazine.

The expedition for the story, which was titled, "First Motor Sortie into Escalante Land," was headed up by writer/photographer Jack Breed. His entourage consisted of a backup support group of fifteen people, a couple of trucks, three Jeeps, and around three dozen horses. They set out to find interesting undocumented natural formations and, fortunately, they found what they were searching for on the very first day.

The colorful red pinnacles and arches were located in an area known locally as "Thorney Pasture." It was used by the area's native population as pasture land for cattle. In honor of the film they were using to photograph the gorgeous scenery, they decided to rename the area "Kodachrome Basin."

Fearing legal repercussions from Kodak, the park was originally named Chimney Rock State Park when it was officially designated a state park in 1962 but was ultimately changed to Kodachrome Basin State Park a few years later with Kodak's blessings.

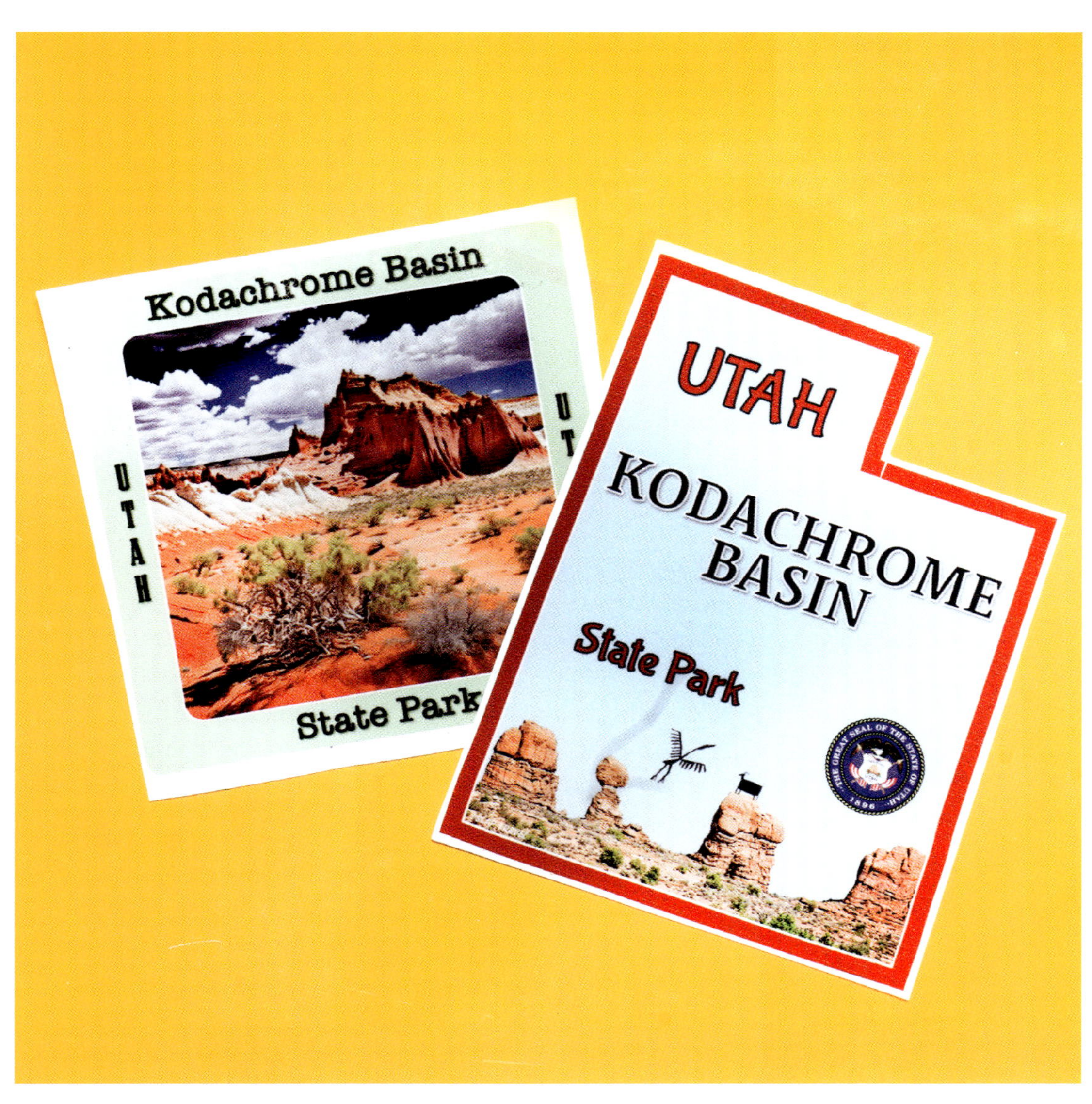
Kodachrome Basin
UTAH
State Park
UTAH
KODACHROME
BASIN
State Park
THE GREAT SEAL OF THE STATE OF UTAH
1896

Kodachrome: The Locomotives

In 1983, a merger was proposed between the Southern Pacific and Atchison, Topeka, and Santa Fe Railway systems, two of the largest West Coast rail freight companies. Formal plans for the consolidation of the two companies was submitted to the Interstate Commerce Commission (ICC) in March 1984, with full confidence the merger would be approved despite push-back from the Justice Department and smaller west coast rail carriers.

Incoming SPSF corporate board members went as far as commissioning a design firm to come up with a color scheme for the new rail system. The proposed design featured a color motif that was highly reminiscent of Kodak's signature red, yellow, and black product packaging. In short time, the newly repainted locomotives and cabooses became known as the "Kodachrome trains." All involved loved the new look and the plan to have the trains re-painted was put into action.

To avoid confusion and any legal problems that might have arisen by using the logo of a yet-to-be- approved corporate merger, half of the locomotives were painted on the sides with the initials "SP" (Southern Pacific), and the other half "SF" (Santa Fe). The initials were positioned in a way that would allow the missing initials to be dropped into place after the merger was finalized. As a teaser, two working locomotives and a caboose were painted with the complete "SPSF" logo.

In July 1986, the ICC decided to put the kibosh on the merger by claiming any public gains suggested by the merger were vastly outweighed by the

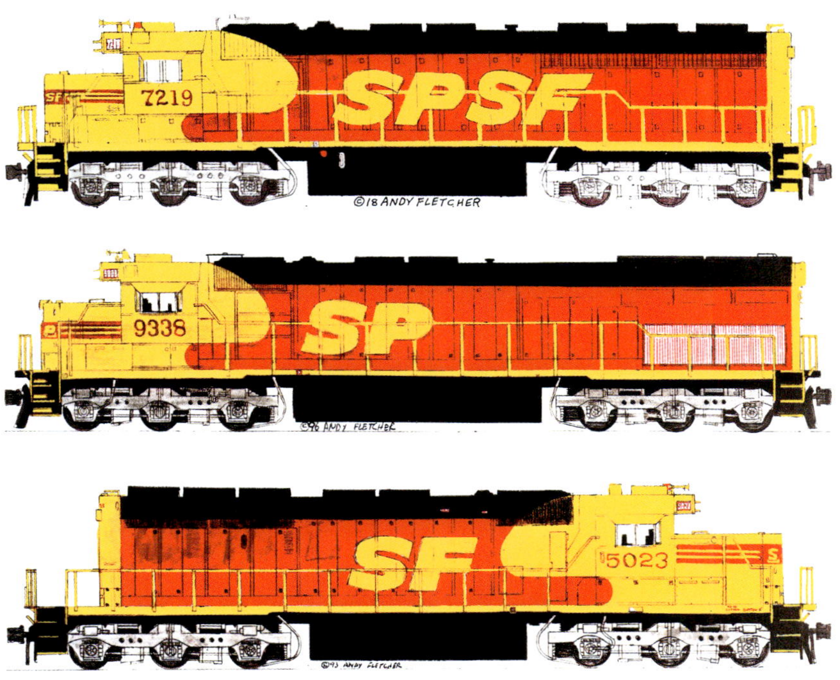

◆ Illustration: Andy Fletcher

negative impact on market prices due to reduced competition among rail lines if the merger were to be approved. As a result, about four hundred locomotives and five cabooses had to be repainted in their original Southern Pacific and Santa Fe color schemes and markings. Within the industry, SPSF quickly became shorthand for "Stop Painting So Fast." By 1990, all traces of the Kodachrome trains were gone. Ironically—yet poetically—what does live on are the hundreds of Kodachrome slides taken of these trains by trainspotters and rail enthusiasts who were active at the time.

Kodachrome Esoterica

Over the years, the Kodachrome logo has appeared on a plethora of promotional products, including thermal picnic coolers, thumb drives, water bottles, mugs, smartphone cases, and pens that double as exposure meters. Here's a sampling.

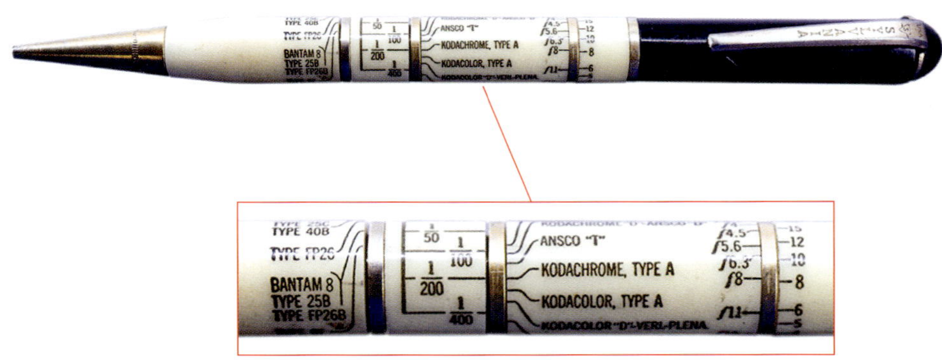

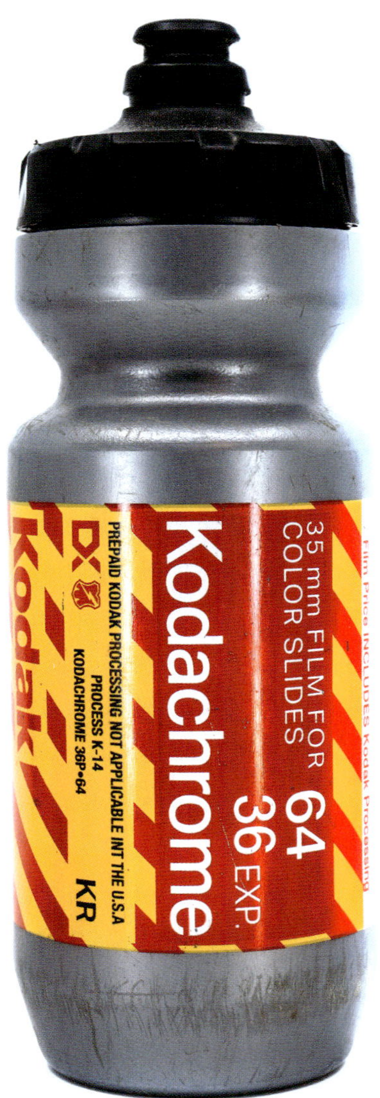
Kodak
DX
PREPAID KODAK PROCESSING NOT APPLICABLE INT THE U.S.A
PROCESS K-14
KODACHROME 36P•64
KR
Kodachrome
35 mm FILM FOR
COLOR SLIDES
64
36 EXP.
Film Price INCLUDES Kodak Processing

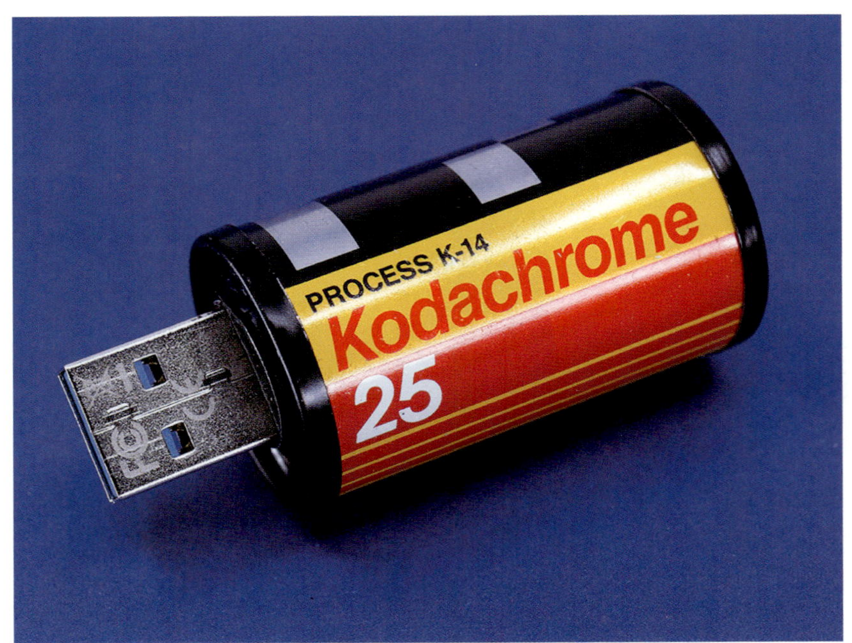

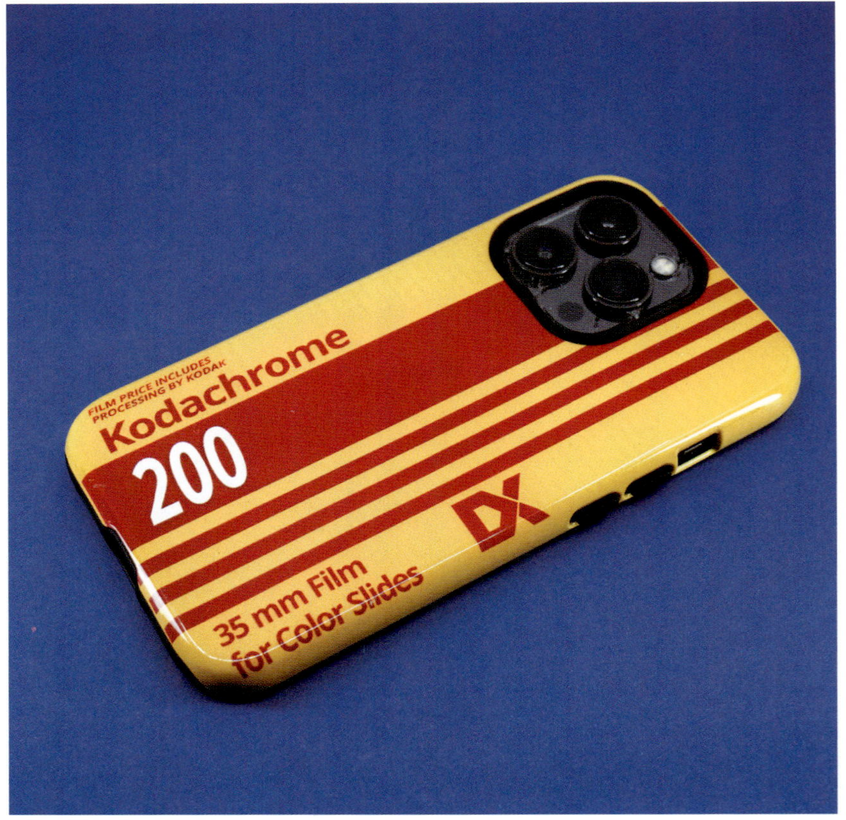

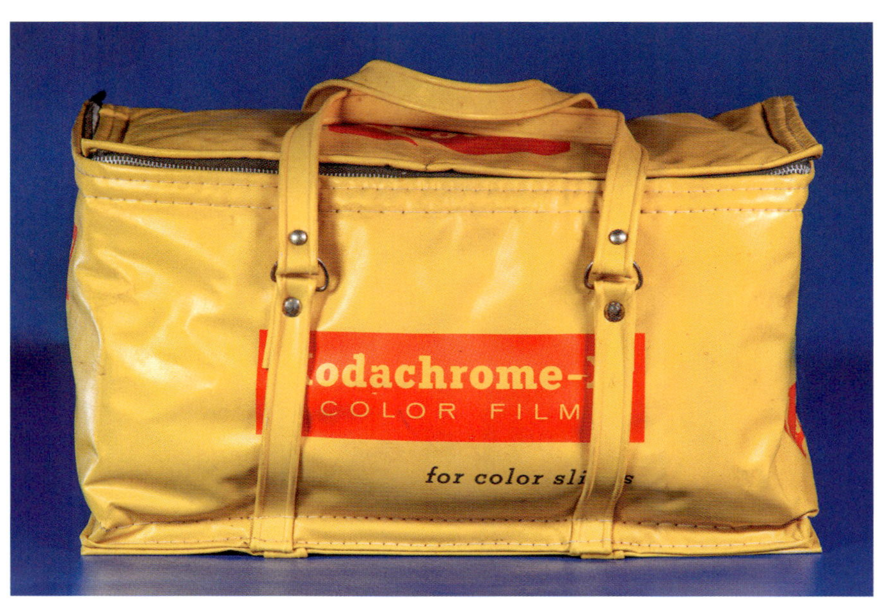

Kodachrome-X
COLOR FILM
for color slides

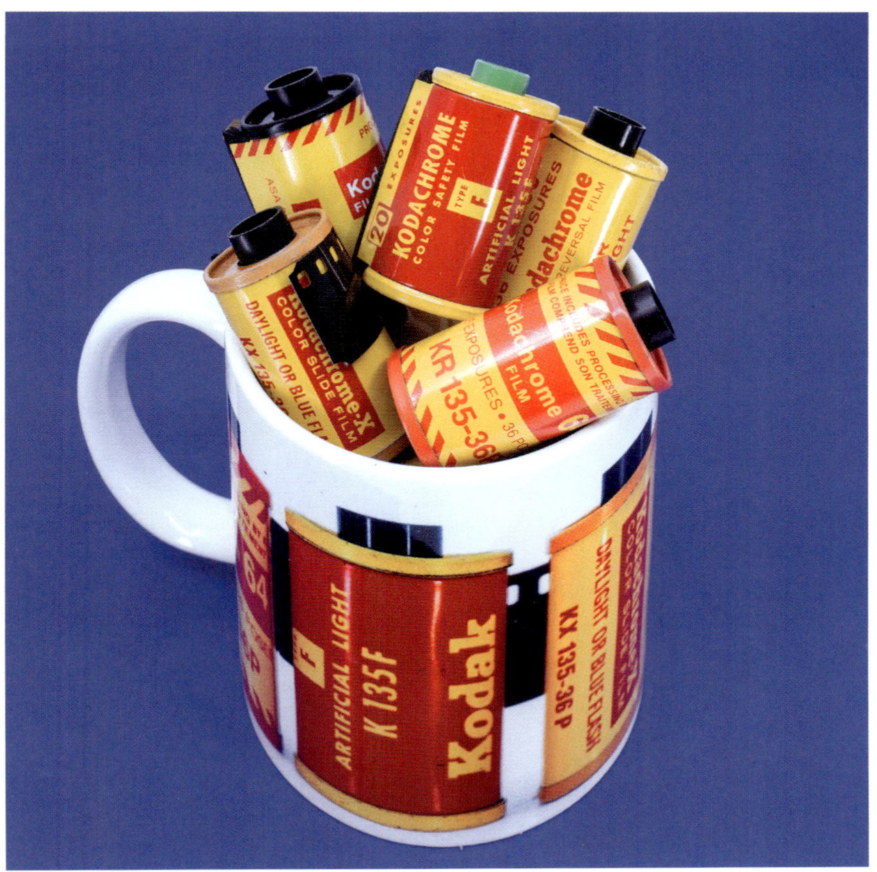

KODACHROME
COLOR SAFETY FILM
K 135
TYPE
F
ARTIFICIAL LIGHT
20 EXPOSURES
Kodachrome
UNIVERSAL FILM
Kodachrome-X
DAYLIGHT OR BLUE FILM
KX 135
KR 135-36
Kodachrome
FILM
ARTIFICIAL LIGHT
K 135F
TYPE
F
Kodak
KODACHROME
COLOR SLIDE
DAYLIGHT OR BLUE FILM
KX 135-36 P

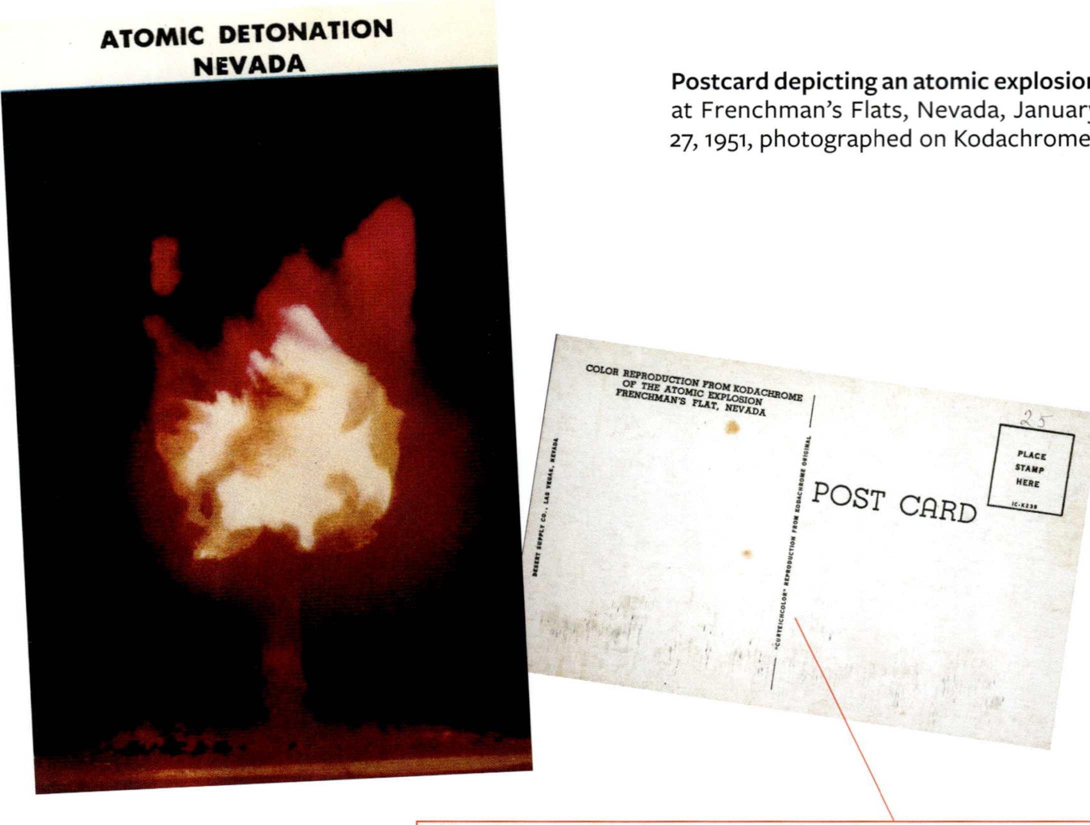

Postcard depicting an atomic explosion at Frenchman's Flats, Nevada, January 27, 1951, photographed on Kodachrome.

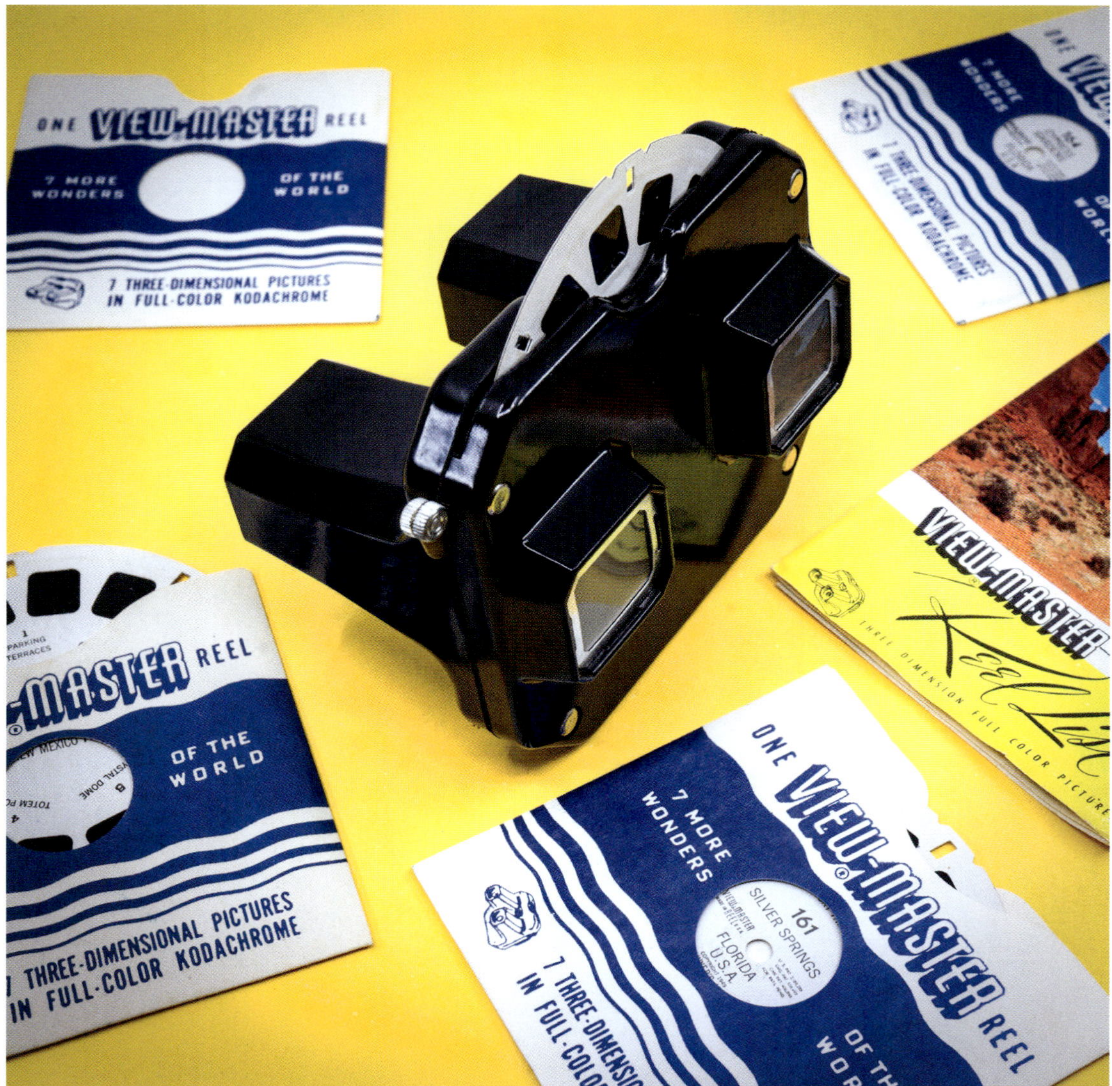

View-Master viewers debuted four years after the introduction of Kodachrome, at the 1939 New York World's Fair.

View-Master circular reels, measuring 3.54 inches in diameter and featuring seven pairs of 12.9 x 11.9mm stereoscopic Kodachrome transparencies circling their edge, quickly gained popularity as three-dimensional alternatives to traditional 2D travel postcards. Production of View- Masters ended in 1998 when the factory was closed following an EPA order due to toxic chemical contamination.

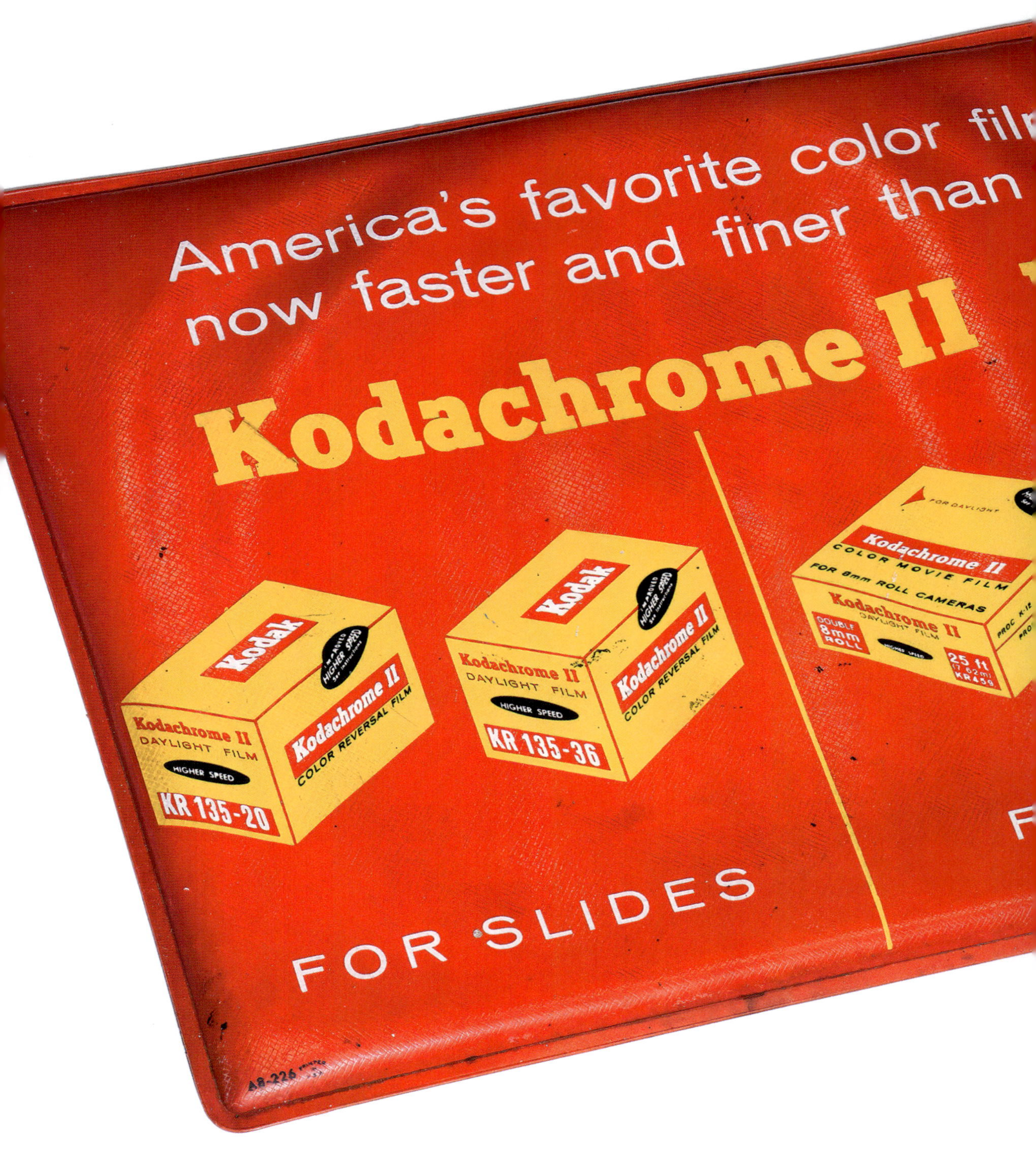
America's favorite color film
now faster and finer than
Kodachrome II
Kodak
Kodachrome II
DAYLIGHT FILM
HIGHER SPEED
KR 135-20
Kodachrome II
DAYLIGHT FILM
HIGHER SPEED
Kodachrome II
COLOR REVERSAL FILM
KR 135-36
FOR DAYLIGHT
Kodachrome II
COLOR MOVIE FILM
FOR 8mm ROLL CAMERAS
Kodachrome II
DAYLIGHT FILM
DOUBLE
8mm
ROLL
25 ft
(7.62m)
KR459
FOR SLIDES

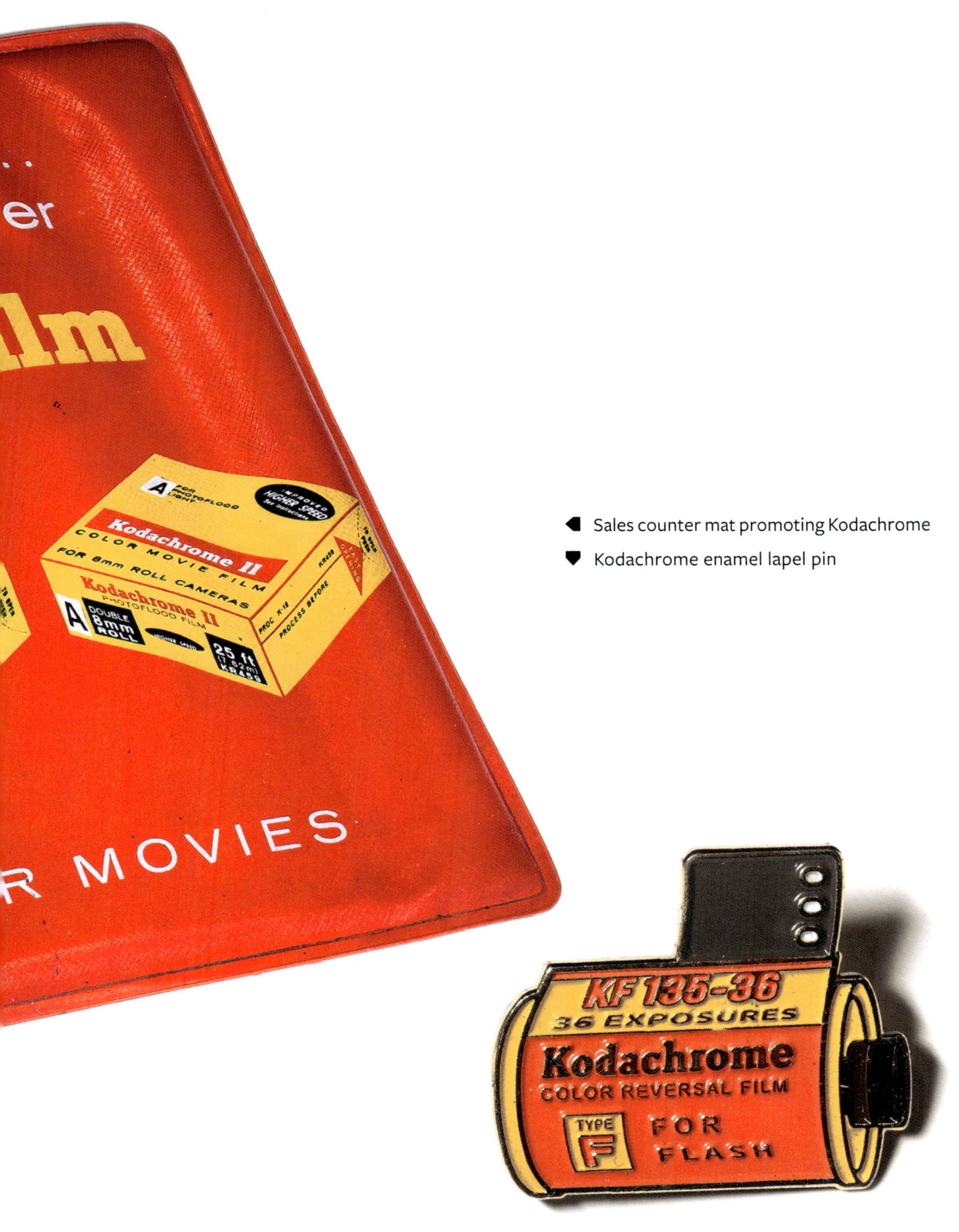

◀ Sales counter mat promoting Kodachrome

▼ Kodachrome enamel lapel pin

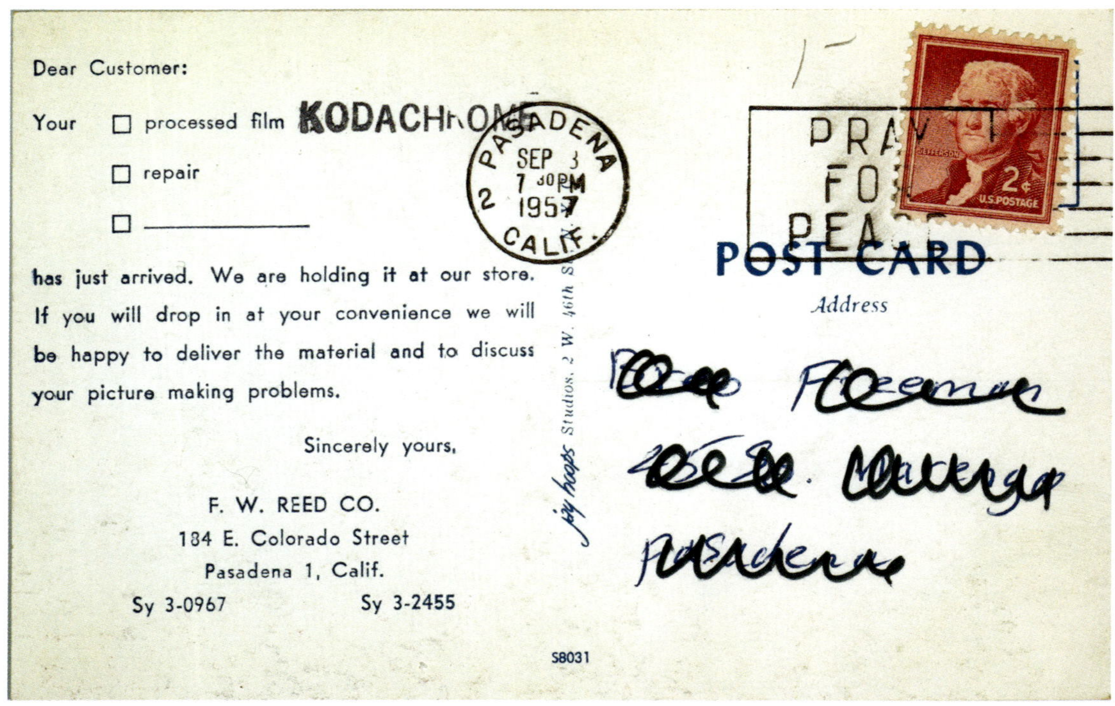

If you dropped off your Kodachrome film for processing at Reed's Camera Store in Pasadena back in the 1950s, they would send you a lovely postcard informing you when it had returned from the local Kodak processing facility.

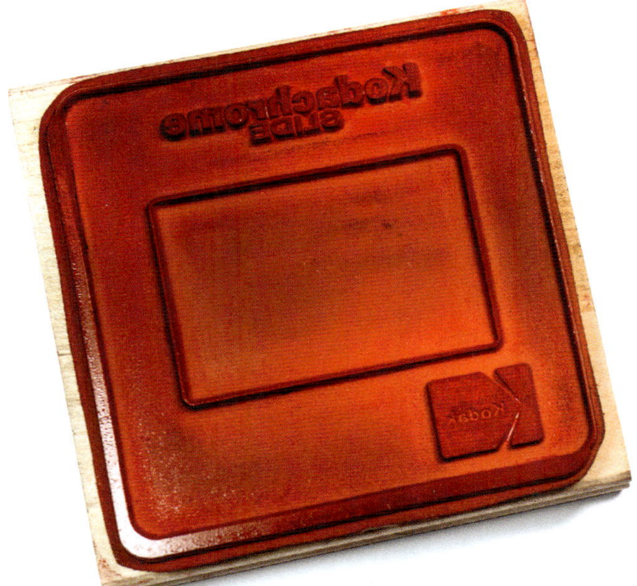

Rubber stamp of a 35mm cardboard
Kodachrome slide mount.

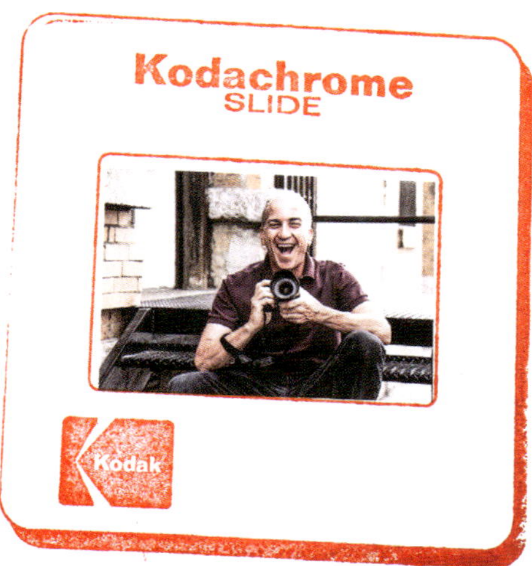

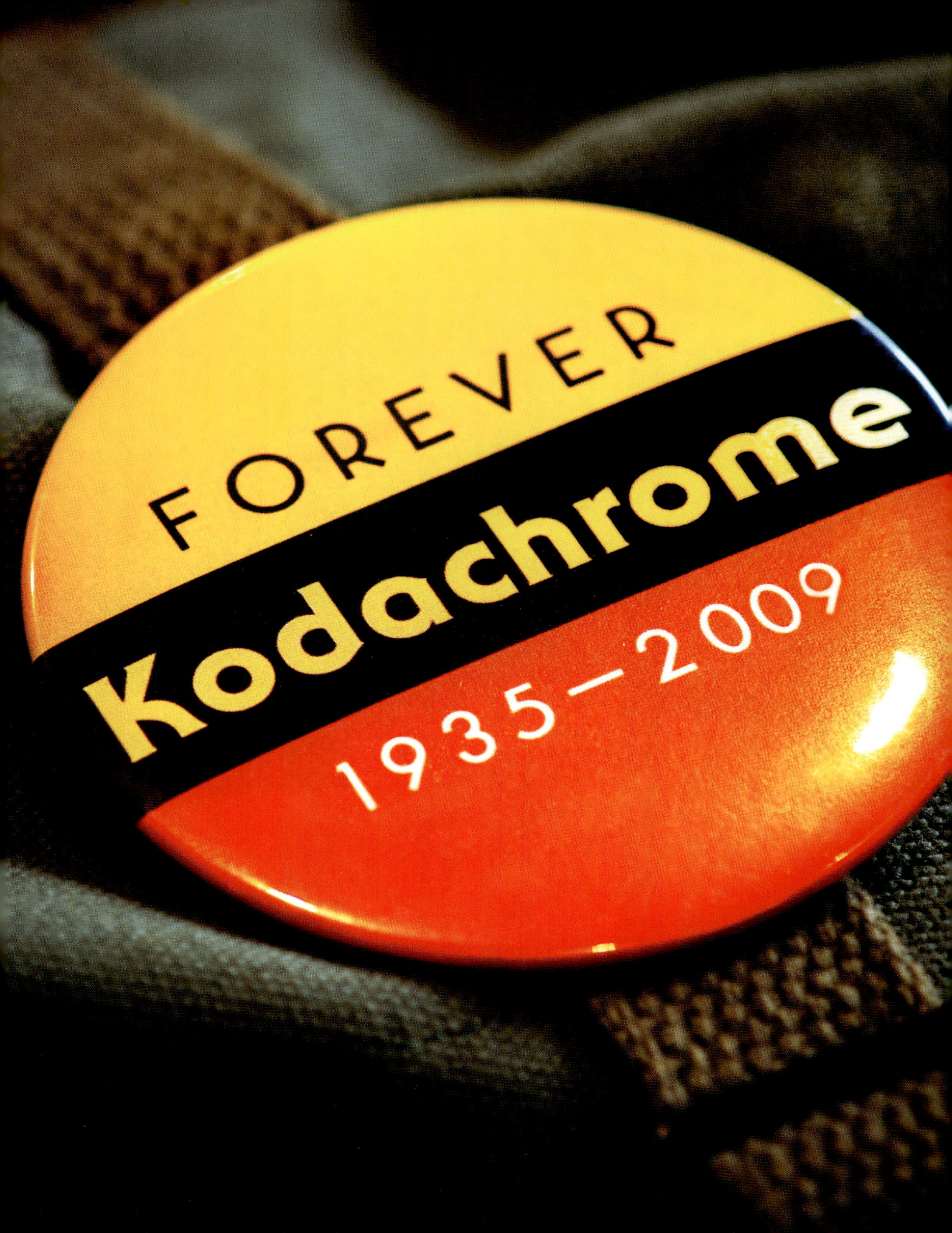
FOREVER
Kodachrome
1935–2009

THE END OF AN ERA

Saying Goodbye to Kodachrome

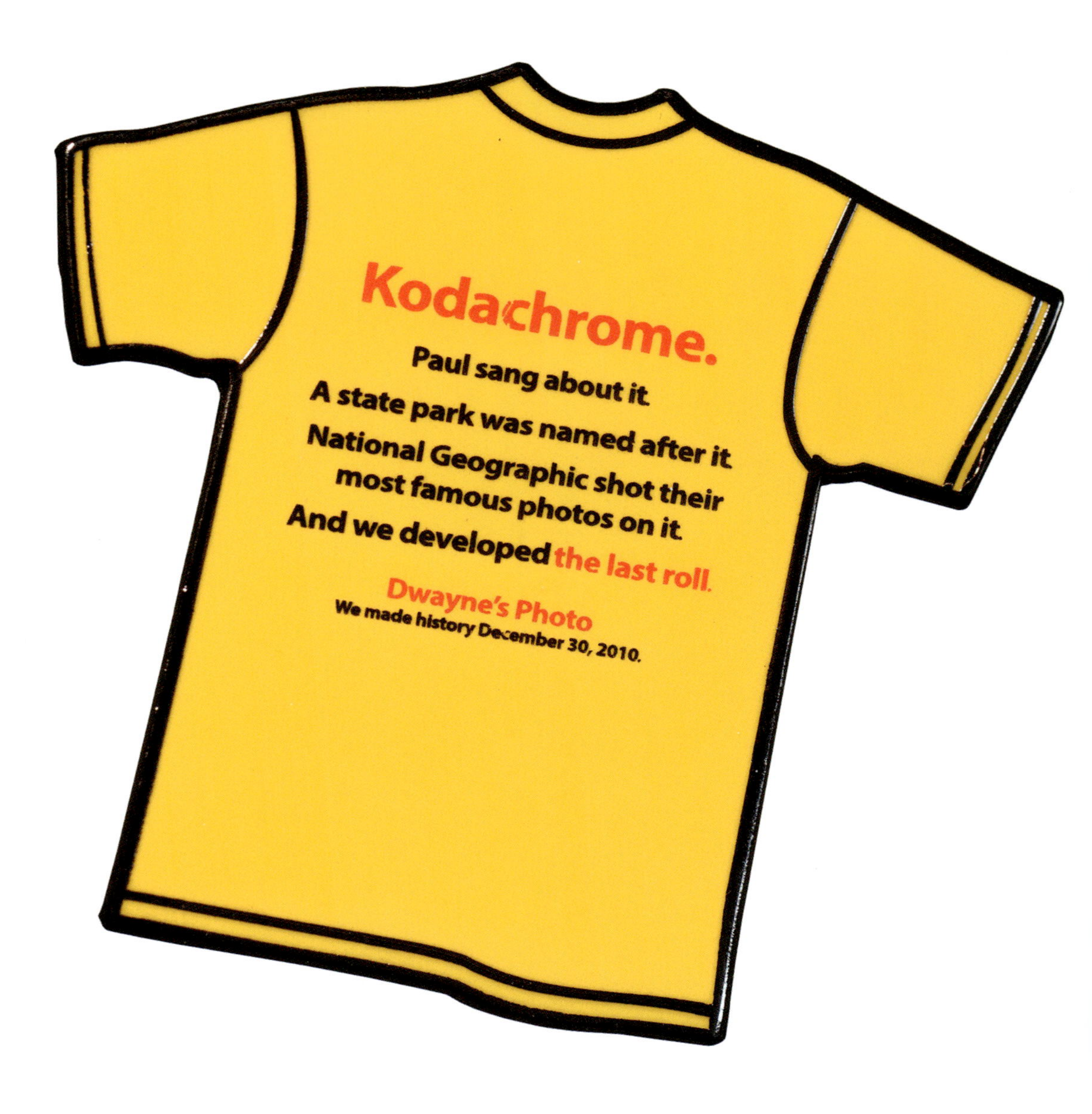
Kodachrome.
Paul sang about it.
A state park was named after it.
National Geographic shot their most famous photos on it.
And we developed the last roll.
Dwayne's Photo
We made history December 30, 2010.

The Final Roll

The last rolls of Kodachrome were scheduled to be processed at Dwayne's Photo in Parsons, Kansas, on December 30, 2010. Kodak had ceased manufacturing the chemistry for processing Kodachrome about a year earlier (June 22, 2009) along with the final production run of Kodachrome 64. But things didn't go as planned.

As word about the deadline neared, Dwayne's was inundated with thousands of rolls of unprocessed Kodachrome. Shortly before the "official" last day, Dwayne Steinle, owner of Dwayne's Photo, decided to make full use of his remaining chemistry and continue processing Kodachrome until the last of the chemistry was gone. The processing line ran for another eight days, with the last rolls exiting the film dryer on January 18, 2011. Included in the final processing run was the very last roll of Kodachrome 64 produced by Kodak.

The honor of exposing that particular roll was granted to Steve McCurry, who took the iconic portrait of Sharbat Gula (better known as the "Afghan Girl") featured on the cover of the June 1985 edition of National Geographic Magazine. The Kodachrome slides McCurry captured with that last roll will be housed in the collection of the George Eastman House, in Rochester, New York.

◀ Dwayne's Photo pin commemorating the
final roll of Kodachrome ever processed

After seventy-four years, Kodachrome was no more. Kodak's Ektachrome and E-6 process reversal films from FUJIFILM and other film manufacturers remain available, but the colors, hues, and archival qualities of Kodachrome are things of the past.

FUJIFILM—of all companies—offers a Kodachrome film simulation called "Classic Chrome" on many of its digital cameras. There are also Lightroom and Photoshop plug-ins that emulate the look of Kodachrome, but even the best emulations are just that—emulations. The colors we see on a computer screen and the colors we see when peering into the highlights and shadows of a Kodachrome slide are totally different animals. Emulations simply cannot capture the punch of bright reds and yellows, let alone the subtleties of a Kodachrome dusk or pre-dawn sky.

As mentioned previously, the photo credits in National Geographic Magazine did not read, "Photograph by..." They read, "Kodachrome by..."

And with that, we bid Godspeed to Kodachrome.

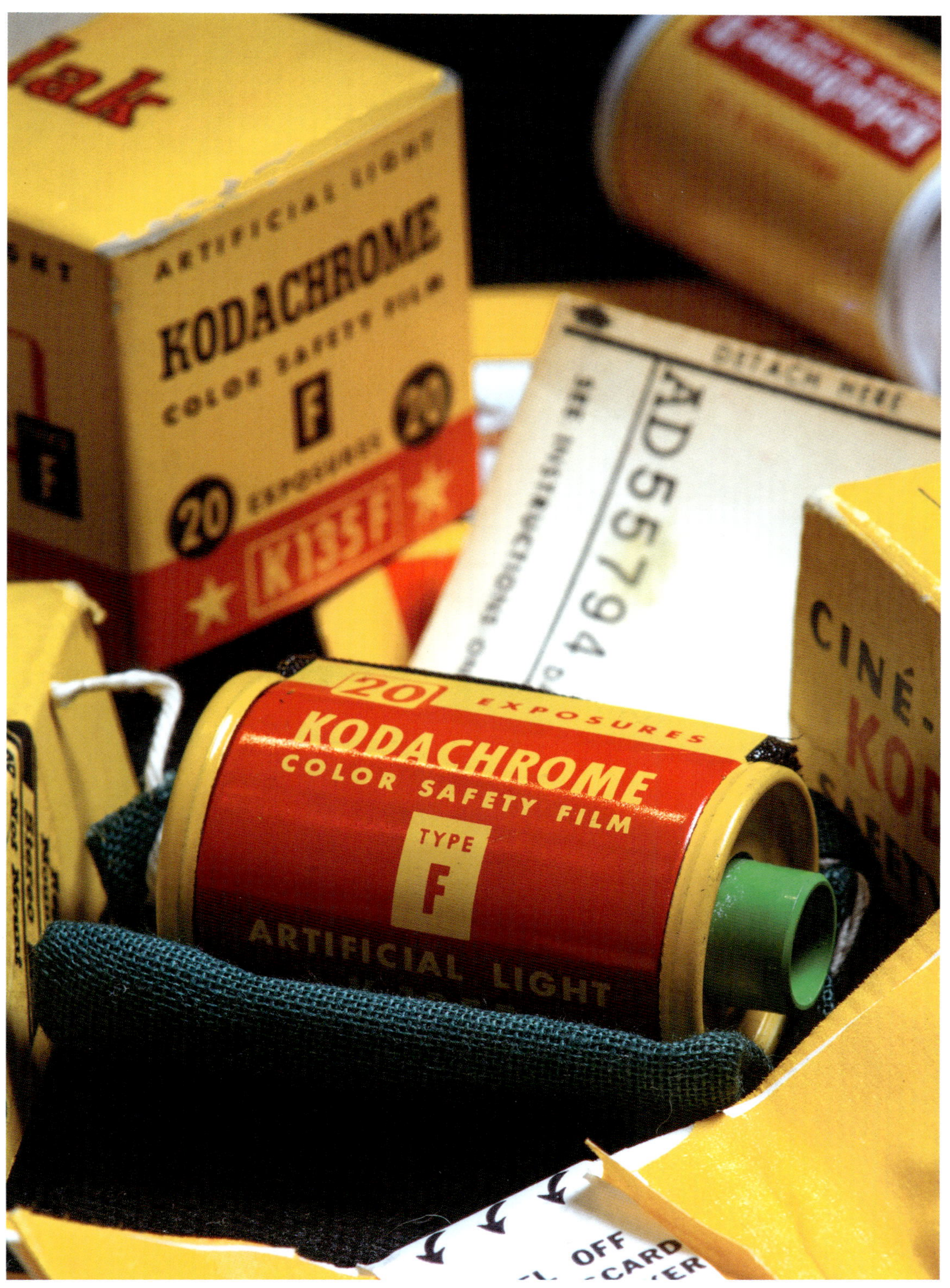
ARTIFICIAL LIGHT
KODACHROME
COLOR SAFETY FILM
F
20 EXPOSURES
K135F
lak
AD 55794
DETACH HERE
SEE INSTRUCTIONS ON
20 EXPOSURES
KODACHROME
COLOR SAFETY FILM
TYPE
F
ARTIFICIAL LIGHT
CINÉ-
KODA
SAFETY
OFF
CARD

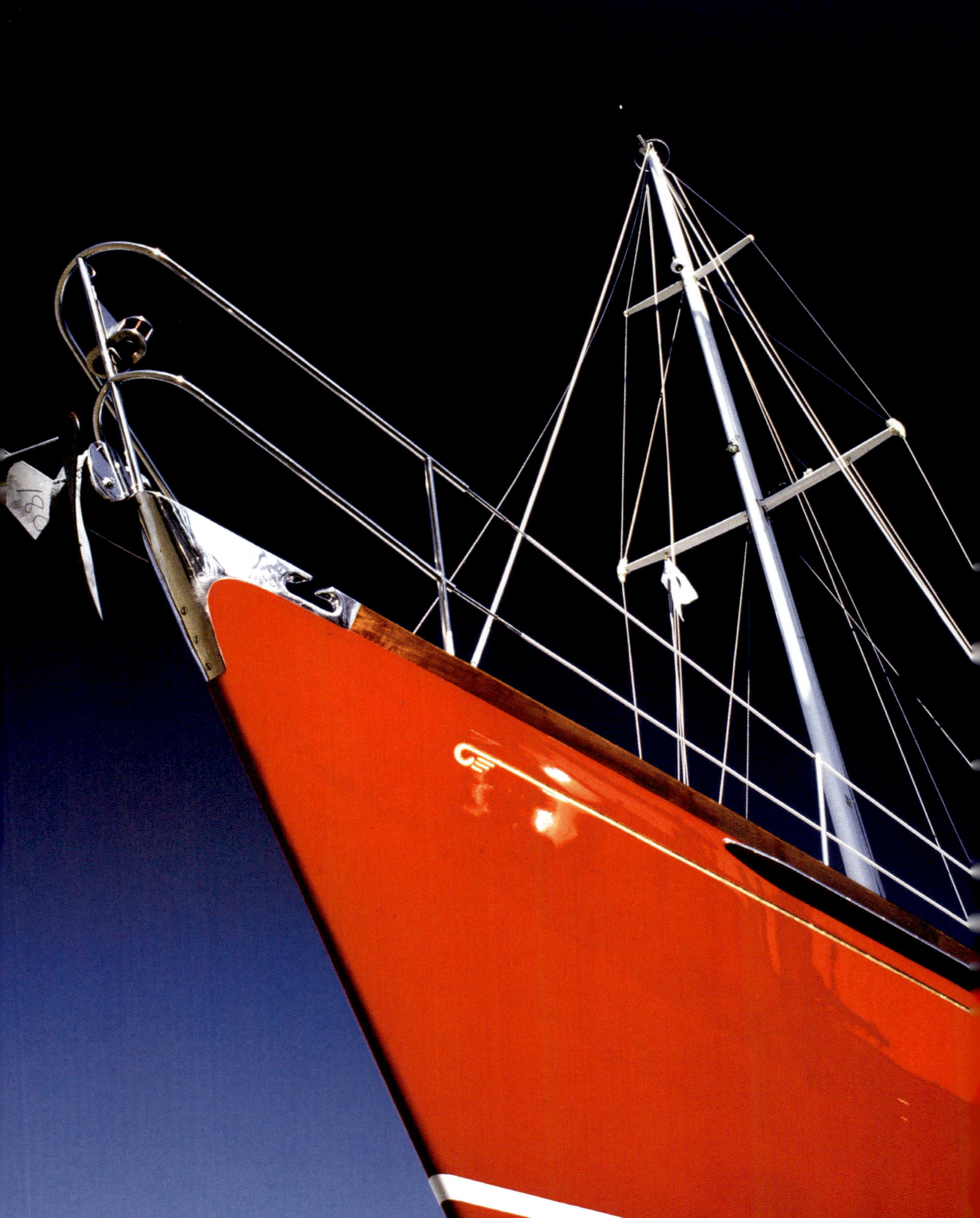

ORIGINAL KODACHROMES

Photographs by Allan Weitz, 1972-1989

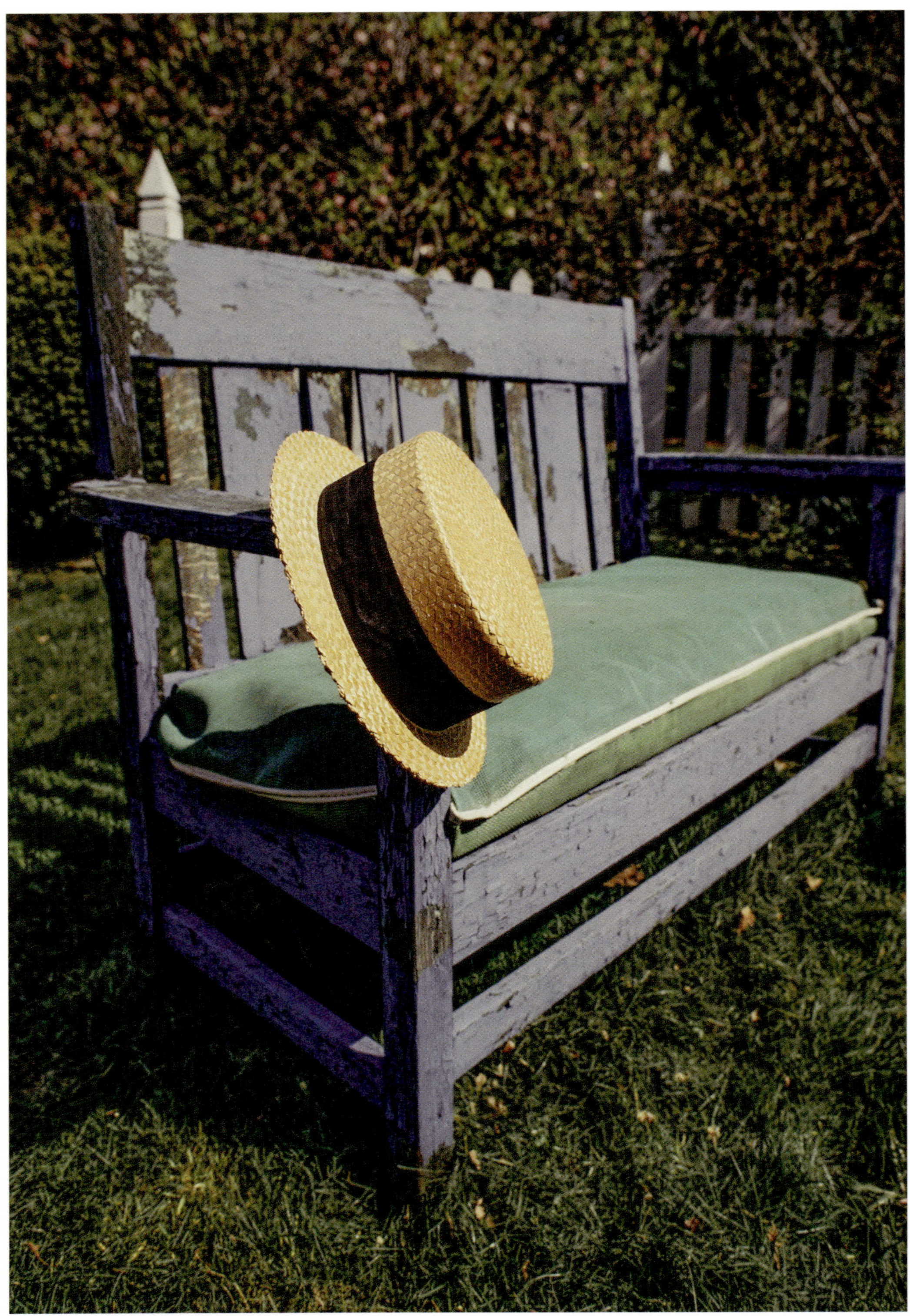

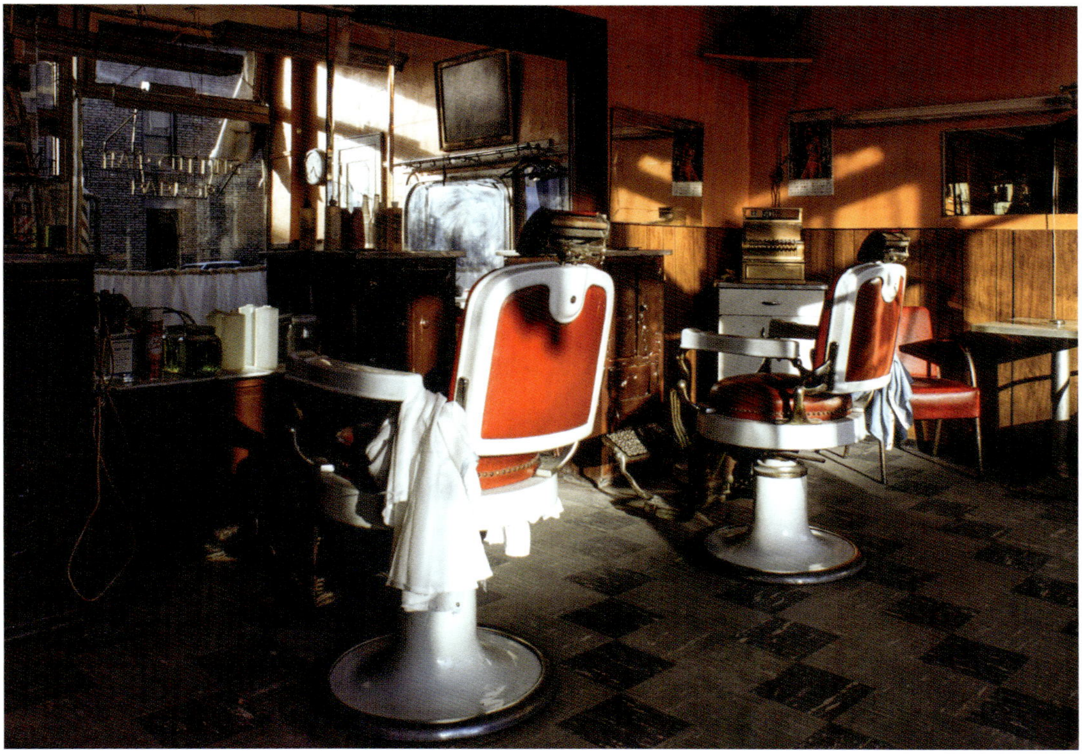

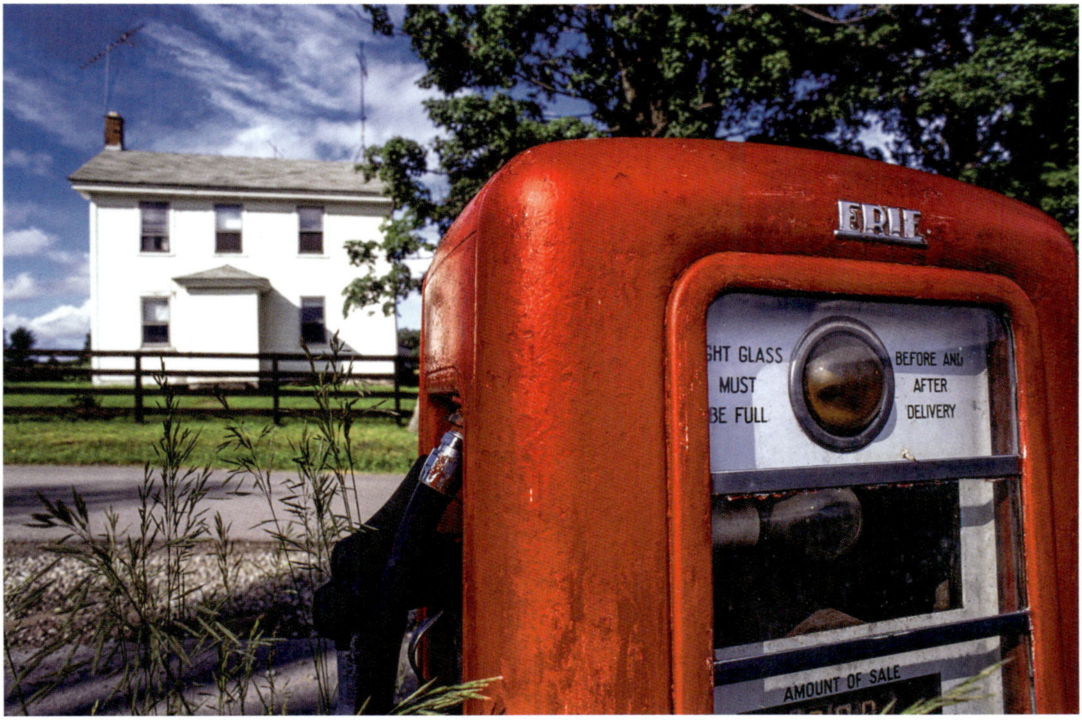
ERIE
GHT GLASS
MUST
BE FULL
BEFORE AND
AFTER
DELIVERY
AMOUNT OF SALE

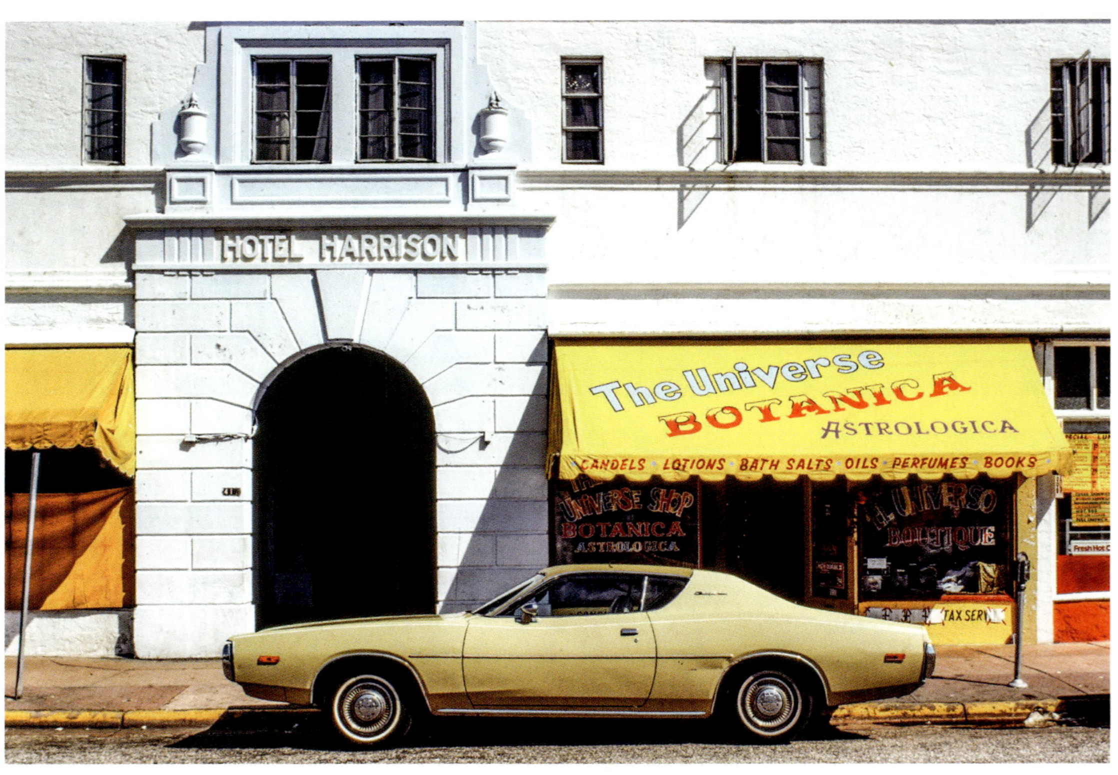
HOTEL HARRISON
The Universe
BOTANICA
ASTROLOGICA
CANDELS · LOTIONS · BATH SALTS · OILS · PERFUMES · BOOKS
UNIVERSE SHOP
BOTANICA
ASTROLOGICA
EL UNIVERSO
BOUTIQUE
TAX SERV
Fresh Hot C

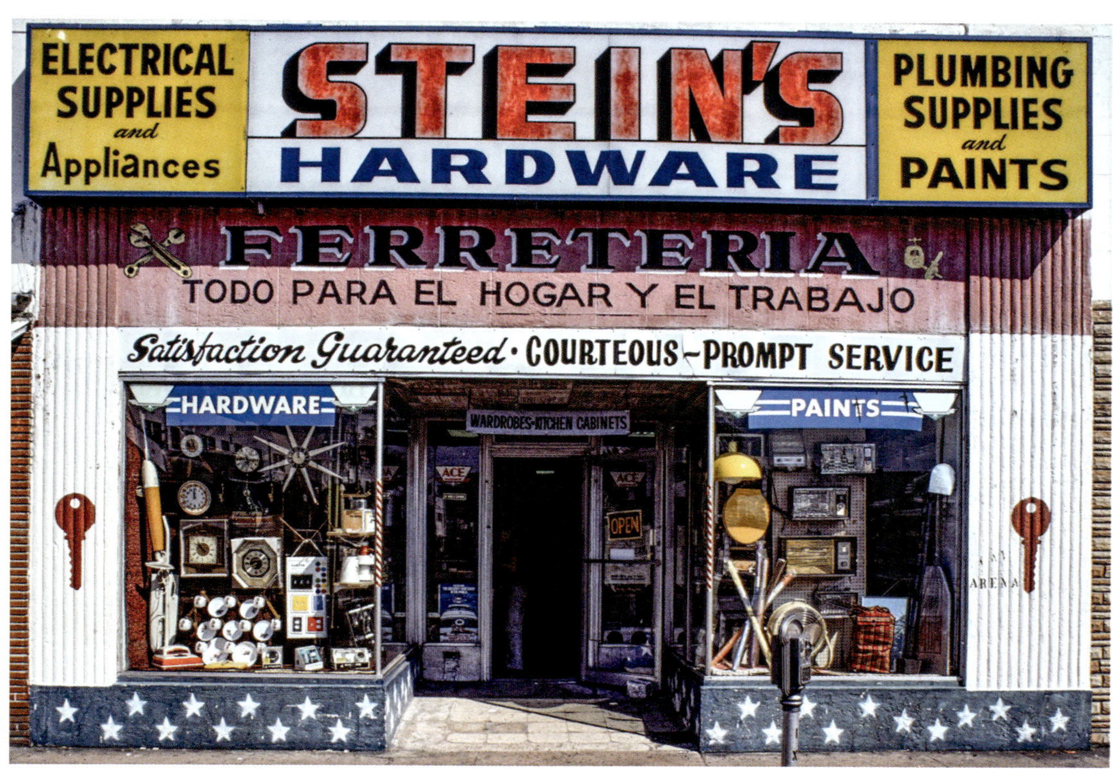

ELECTRICAL SUPPLIES
and
Appliances
STEIN'S
HARDWARE
PLUMBING SUPPLIES
and
PAINTS
FERRETERIA
TODO PARA EL HOGAR Y EL TRABAJO
Satisfaction Guaranteed · COURTEOUS - PROMPT SERVICE
HARDWARE
PAINTS
WARDROBES-KITCHEN CABINETS
ACE
ACE
OPEN

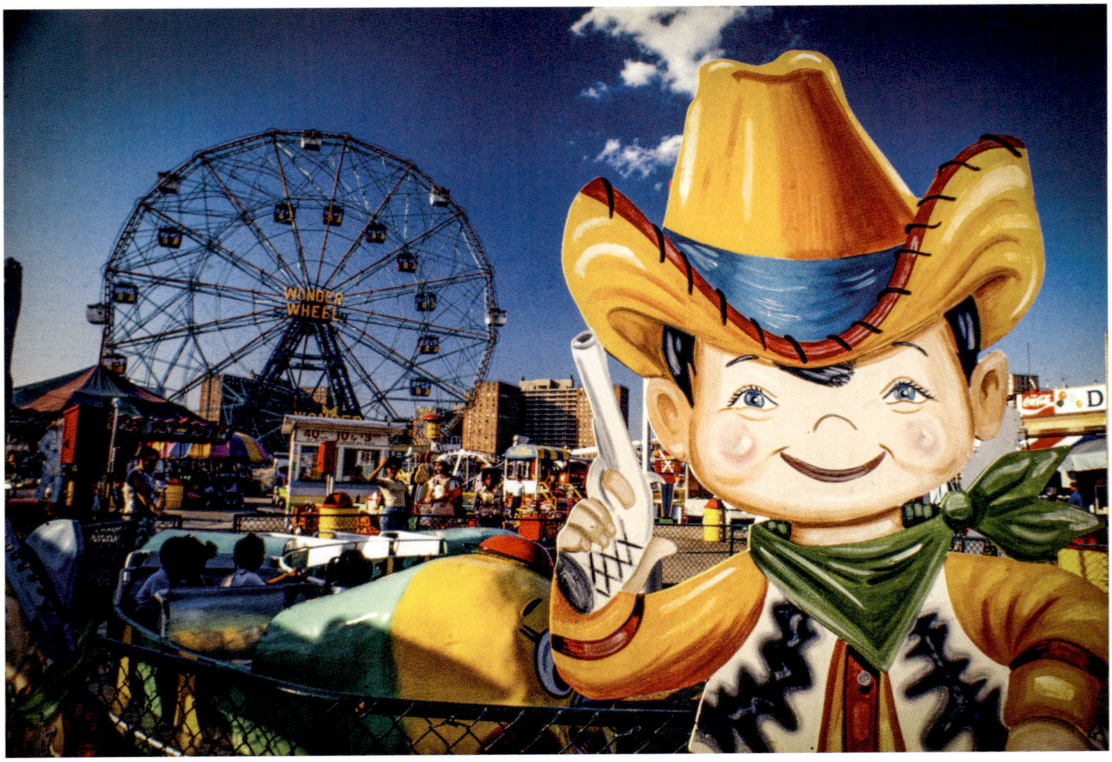

WONDER
WHEEL

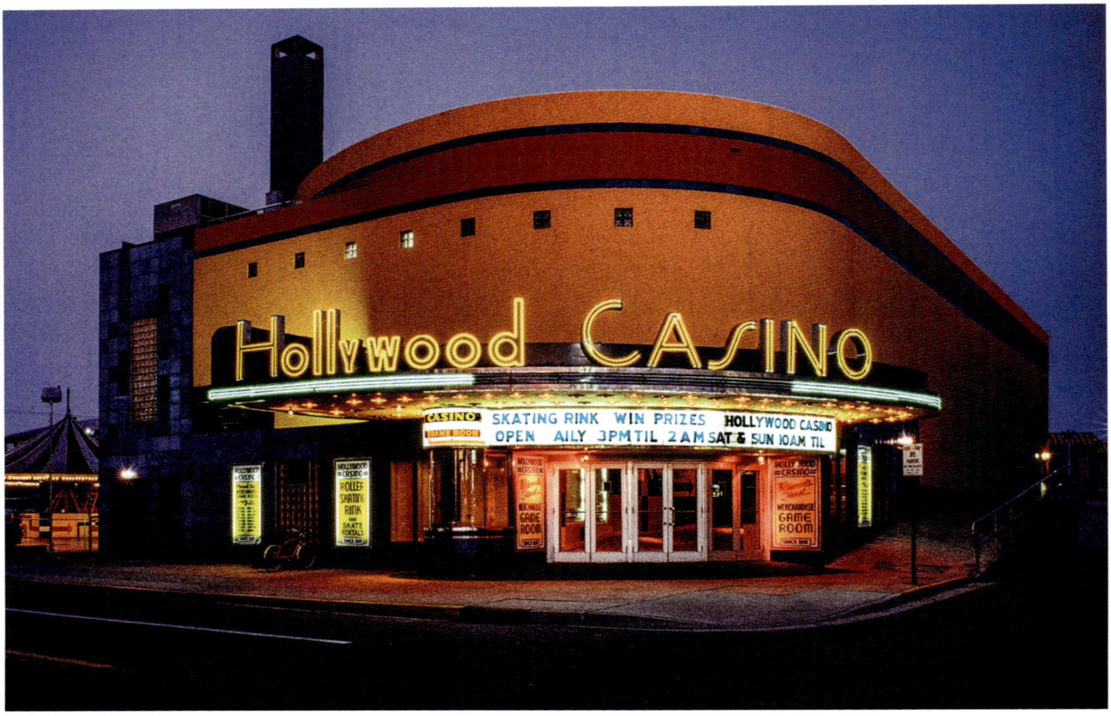

Hollywood CASINO
CASINO
GAME ROOM
SKATING RINK   WIN PRIZES   HOLLYWOOD CASINO
OPEN   AILY 3PM TIL 2 AM SAT & SUN 10AM TIL
ROLLER
SKATING
RINK
AND
SKATE
RENTALS
MERCHANDISE
GAME
ROOM
MERCHANDISE
GAME
ROOM

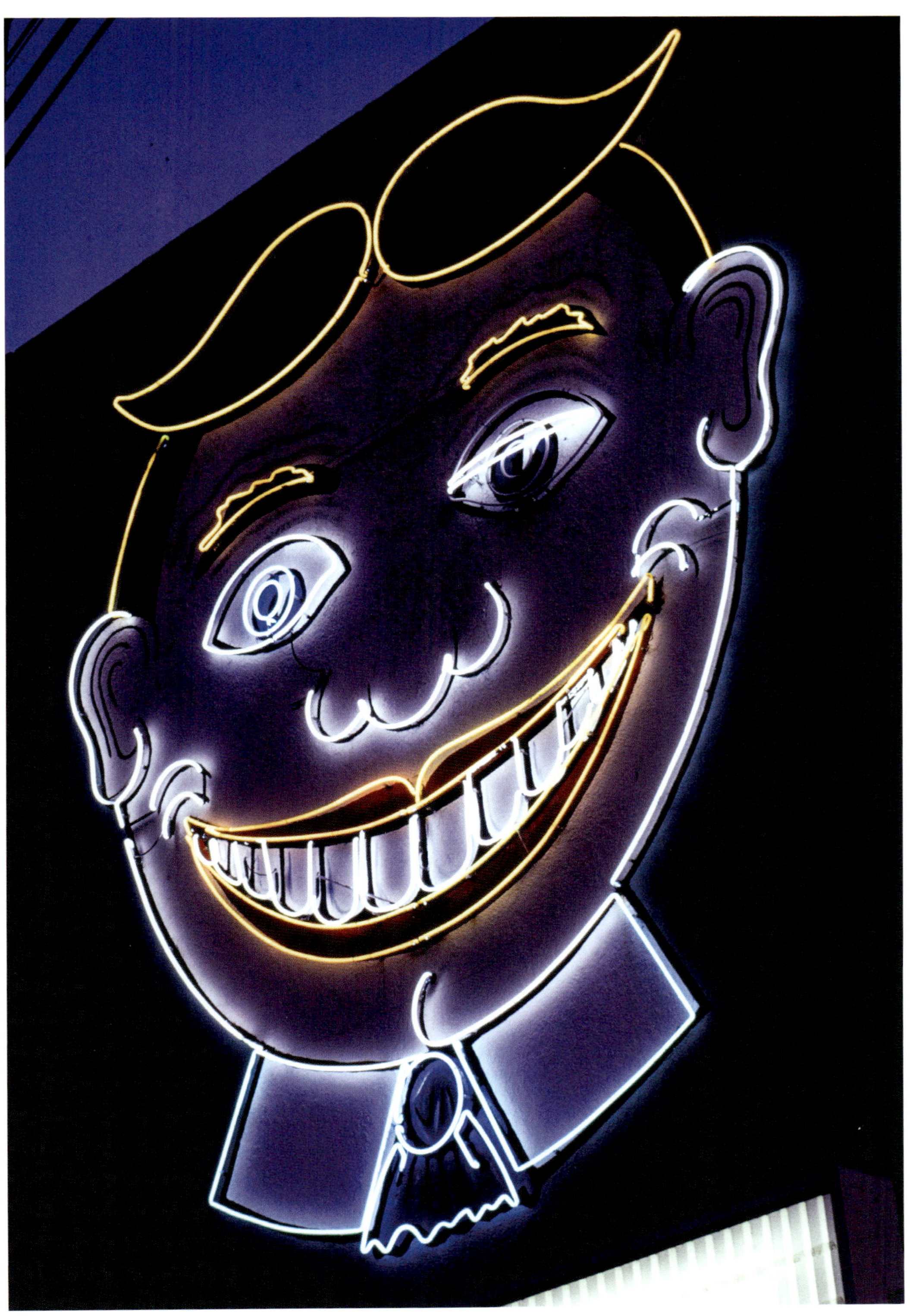

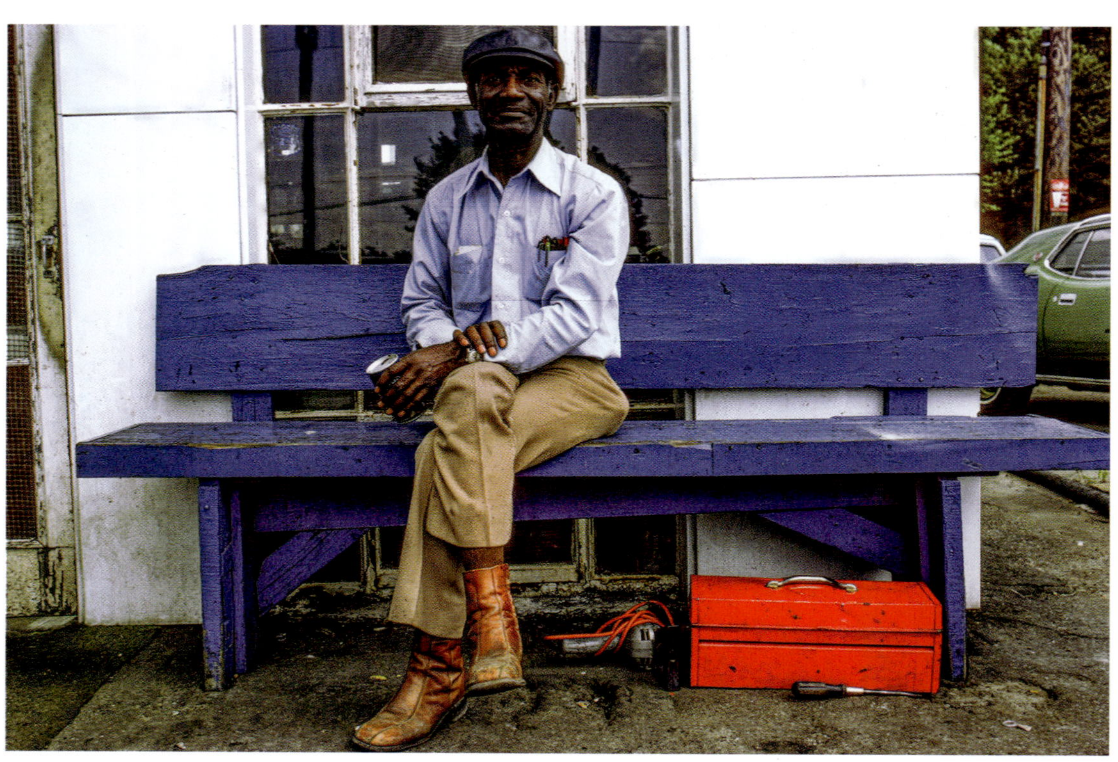

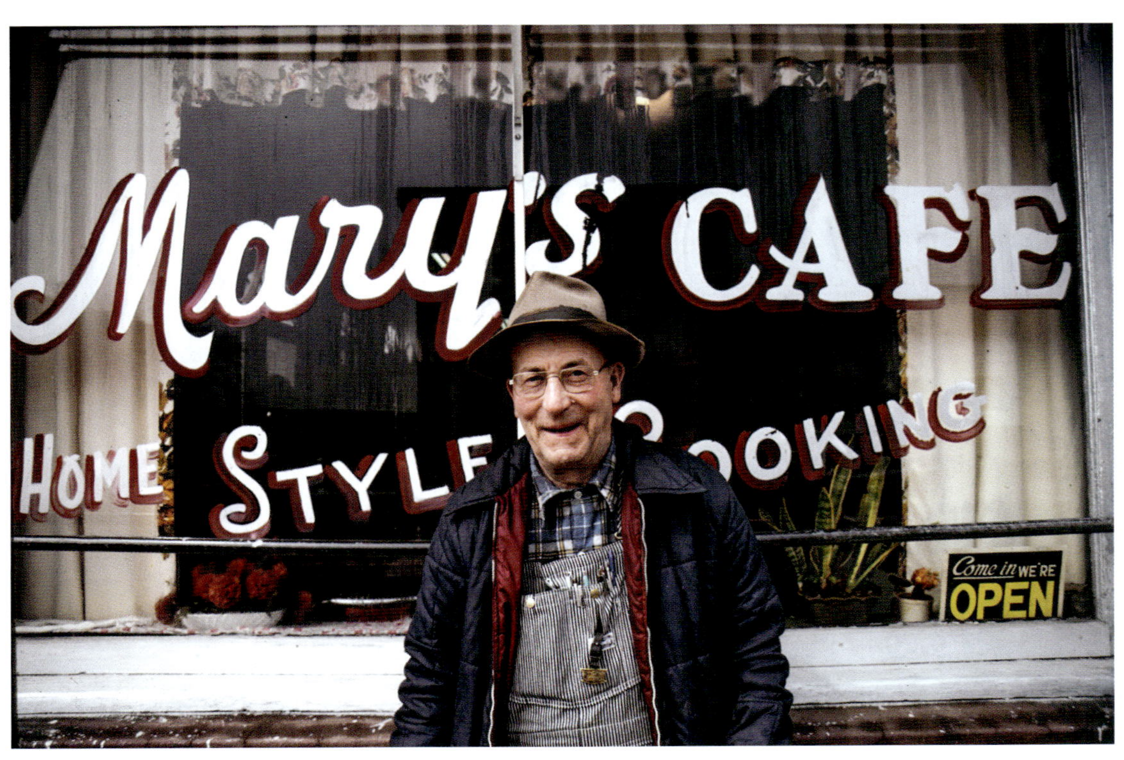
Mary's CAFE
HOME STYLE COOKING
Come in WE'RE
OPEN

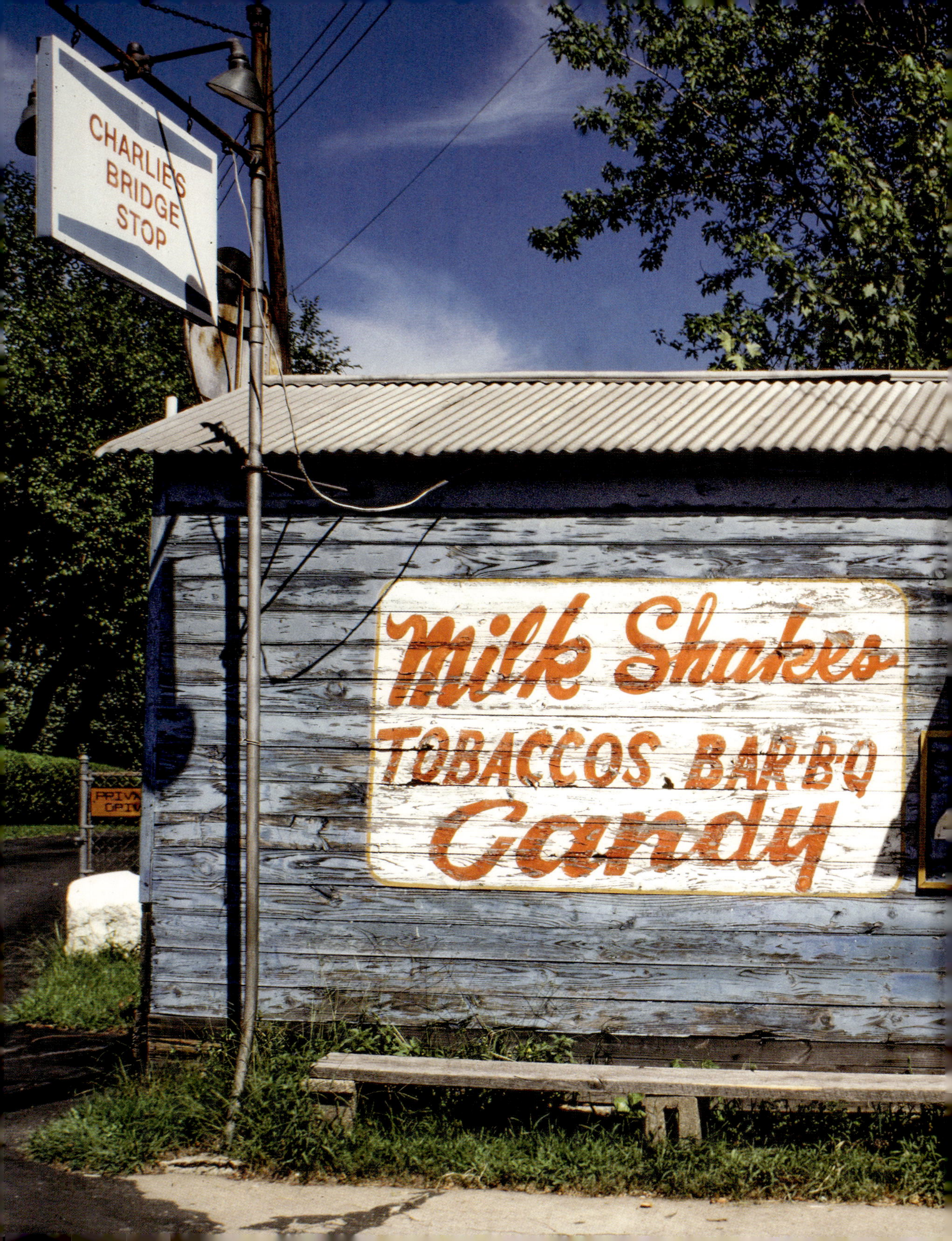

CHARLIE'S
BRIDGE
STOP

Milk Shakes

TOBACCOS. BAR BQ

Candy

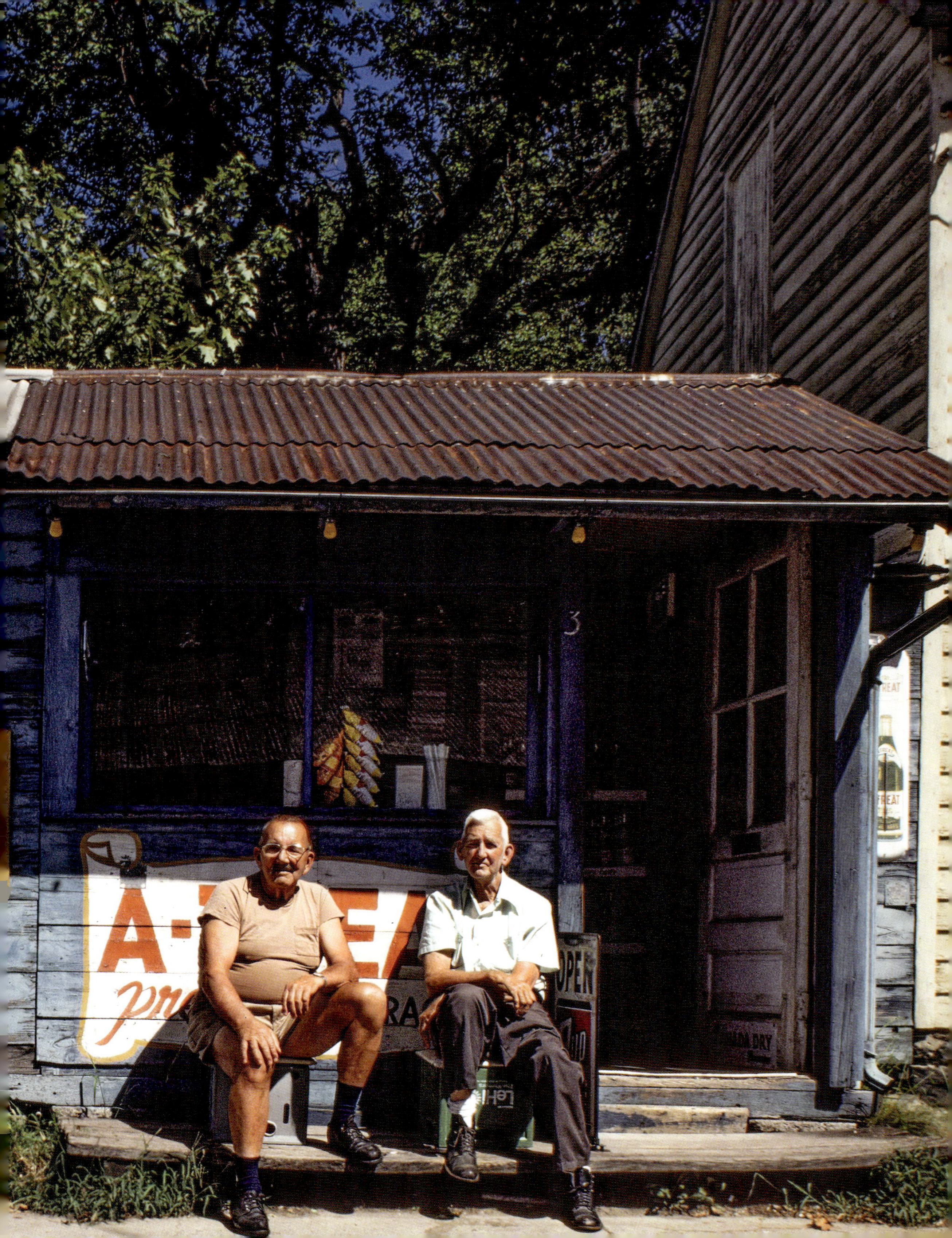
A-T EZ
pr
OPEN
LEHIGH
TREAT
DRY

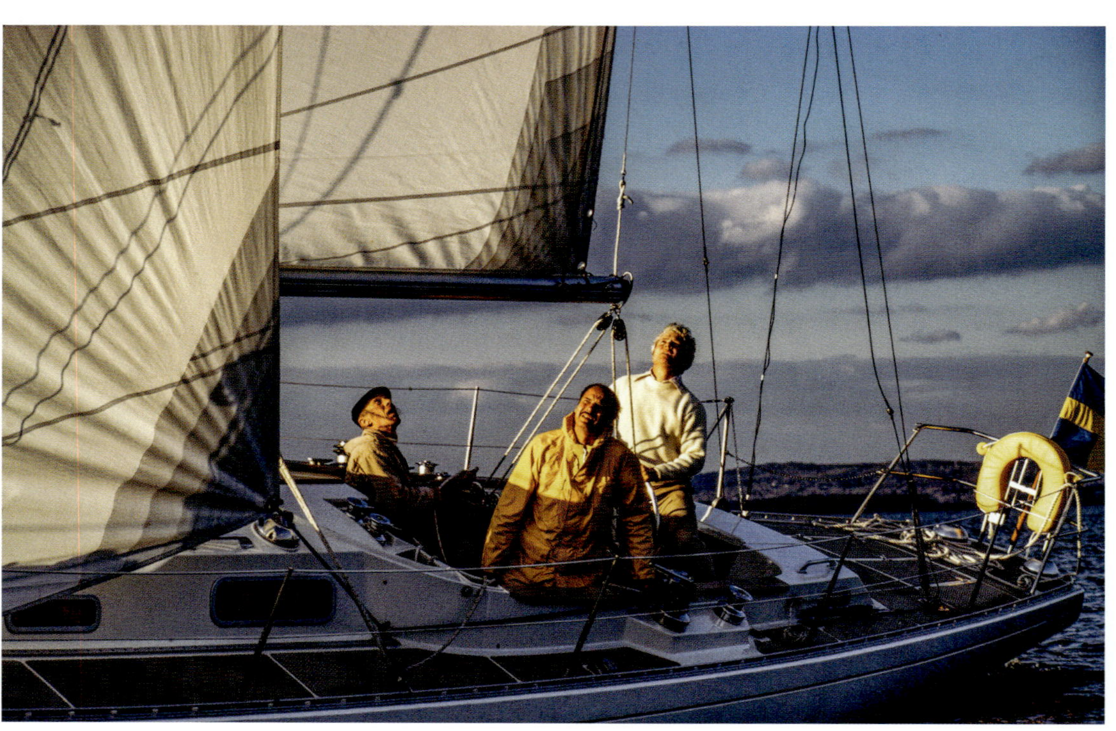

 Original Kodachromes

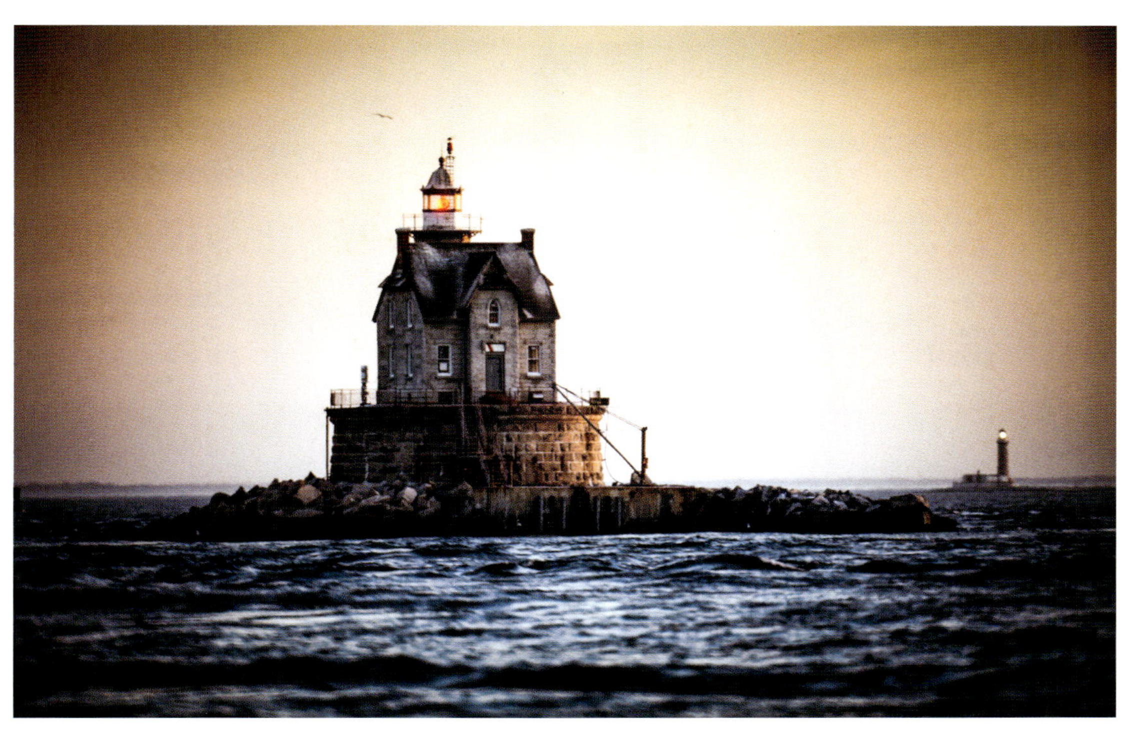

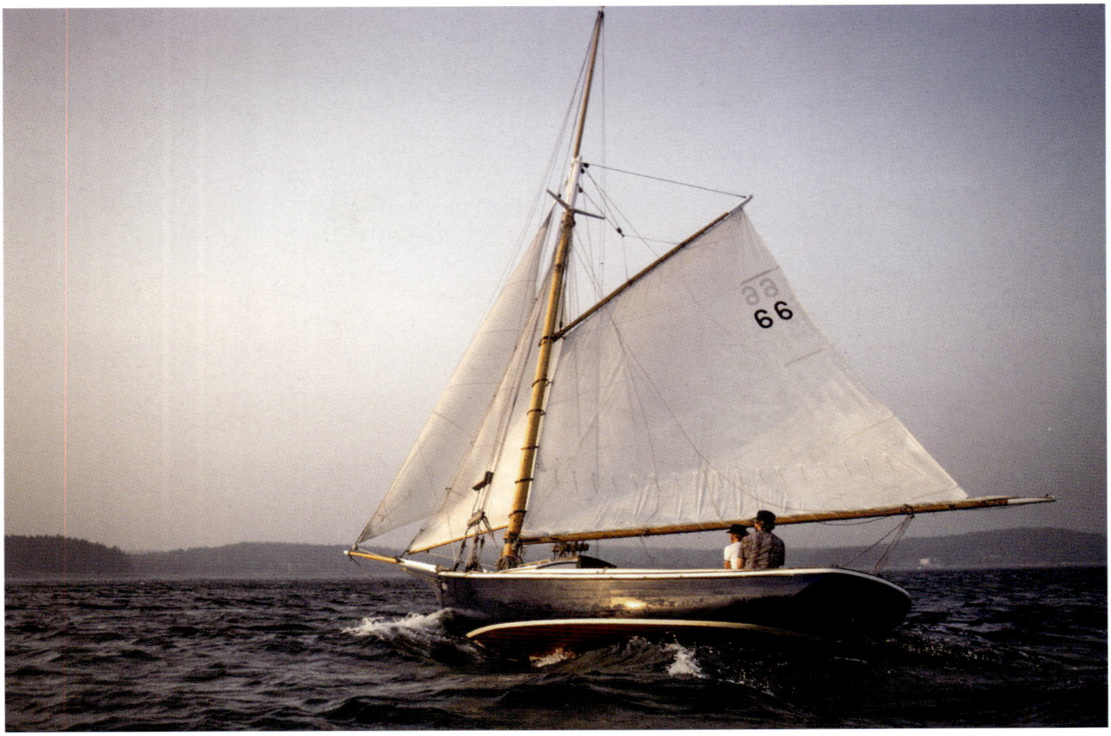

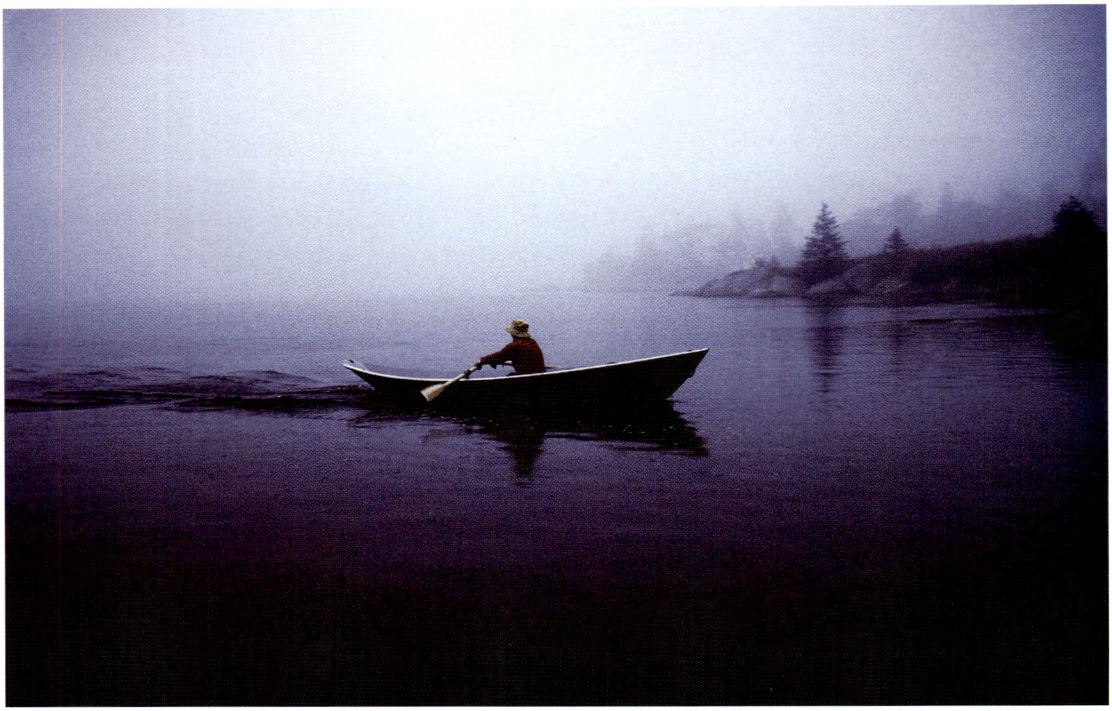

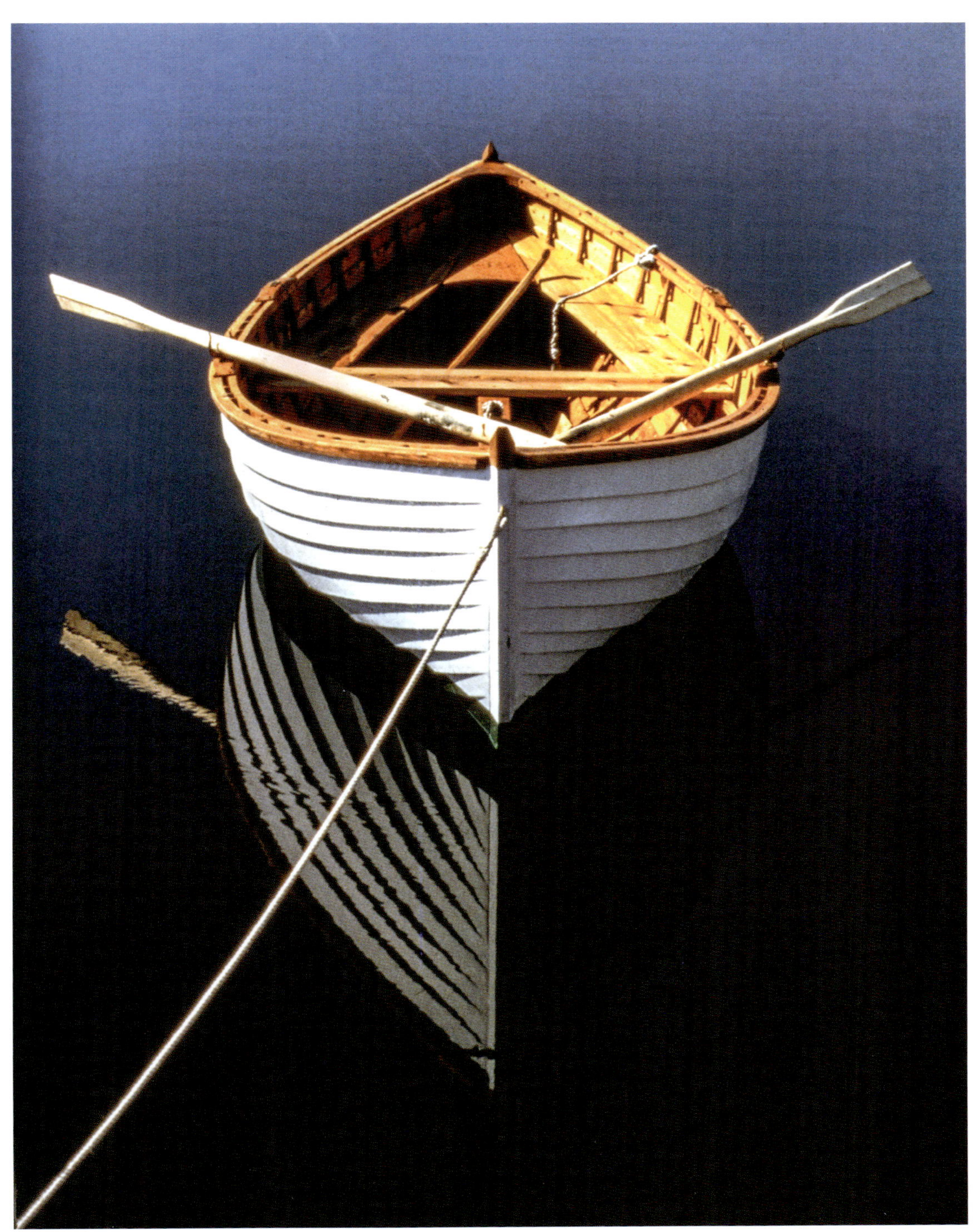

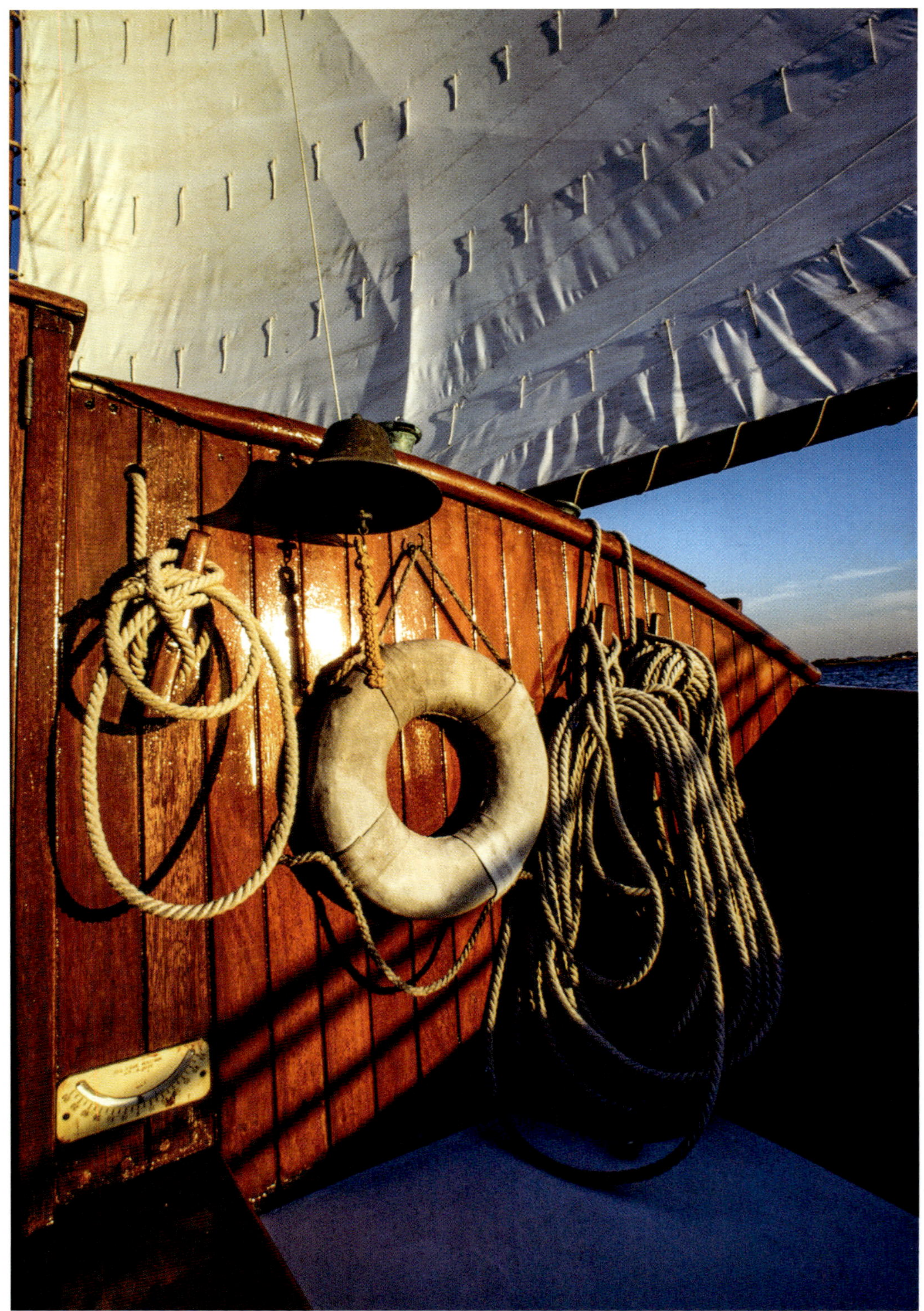

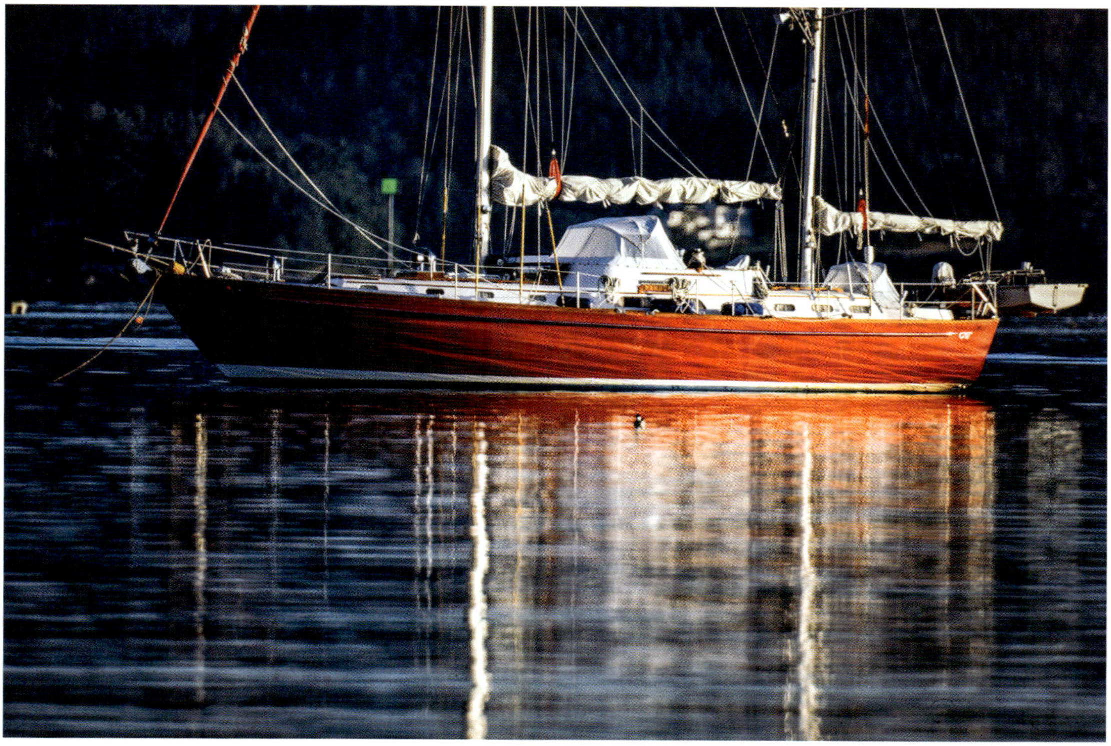

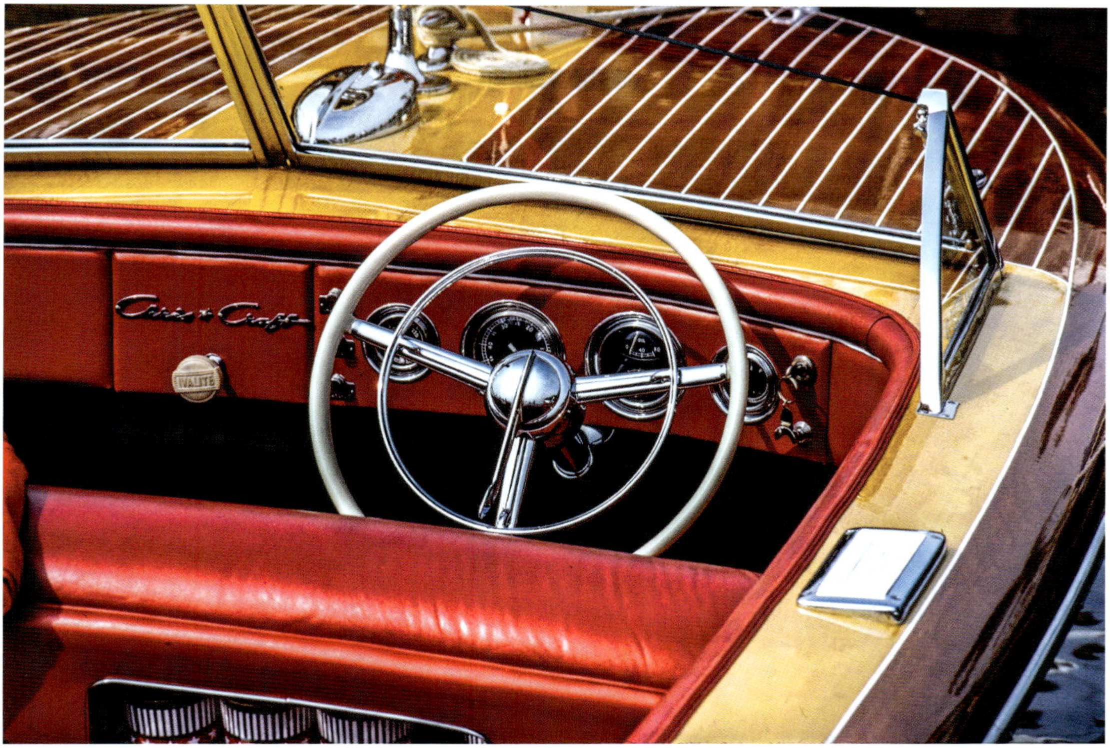

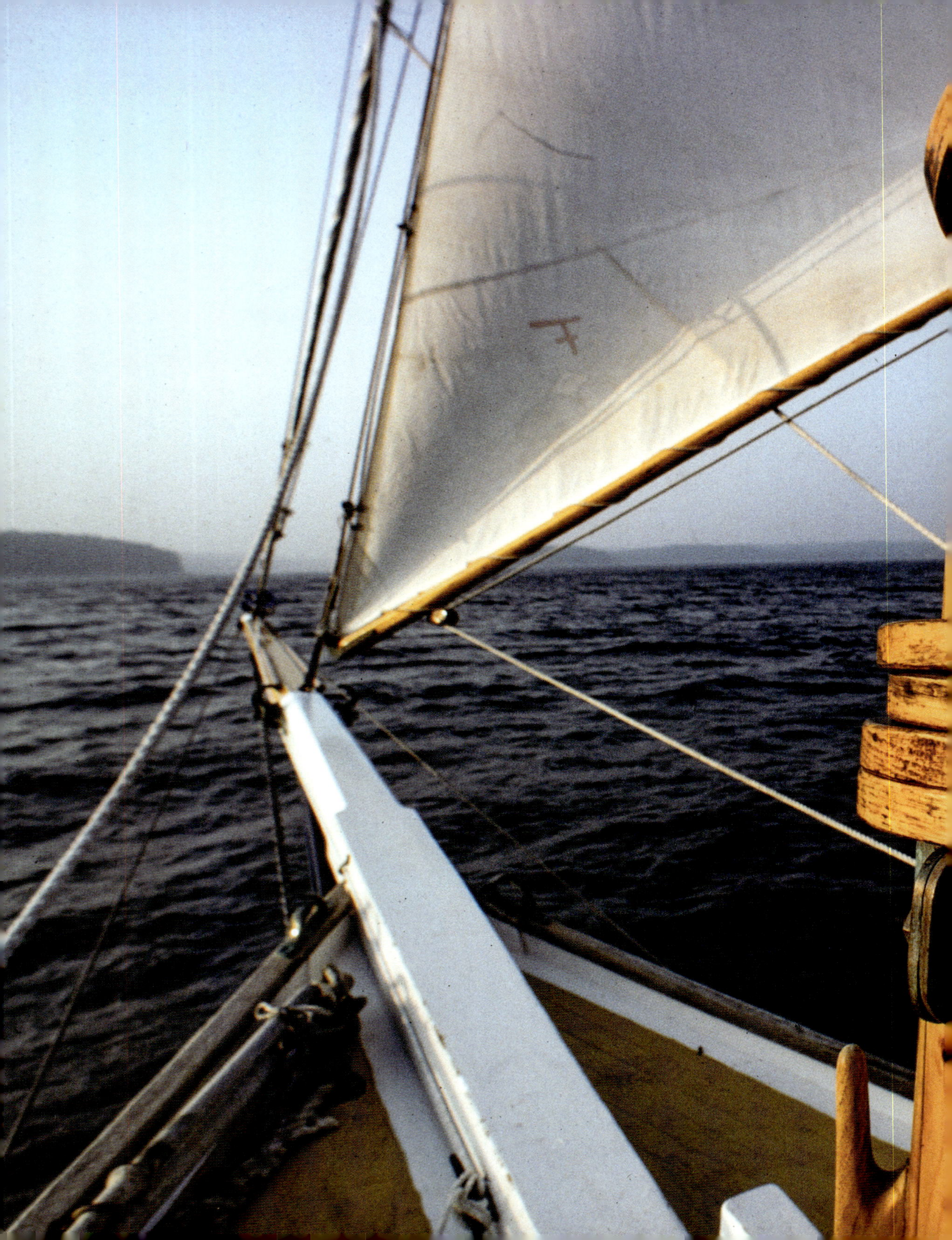

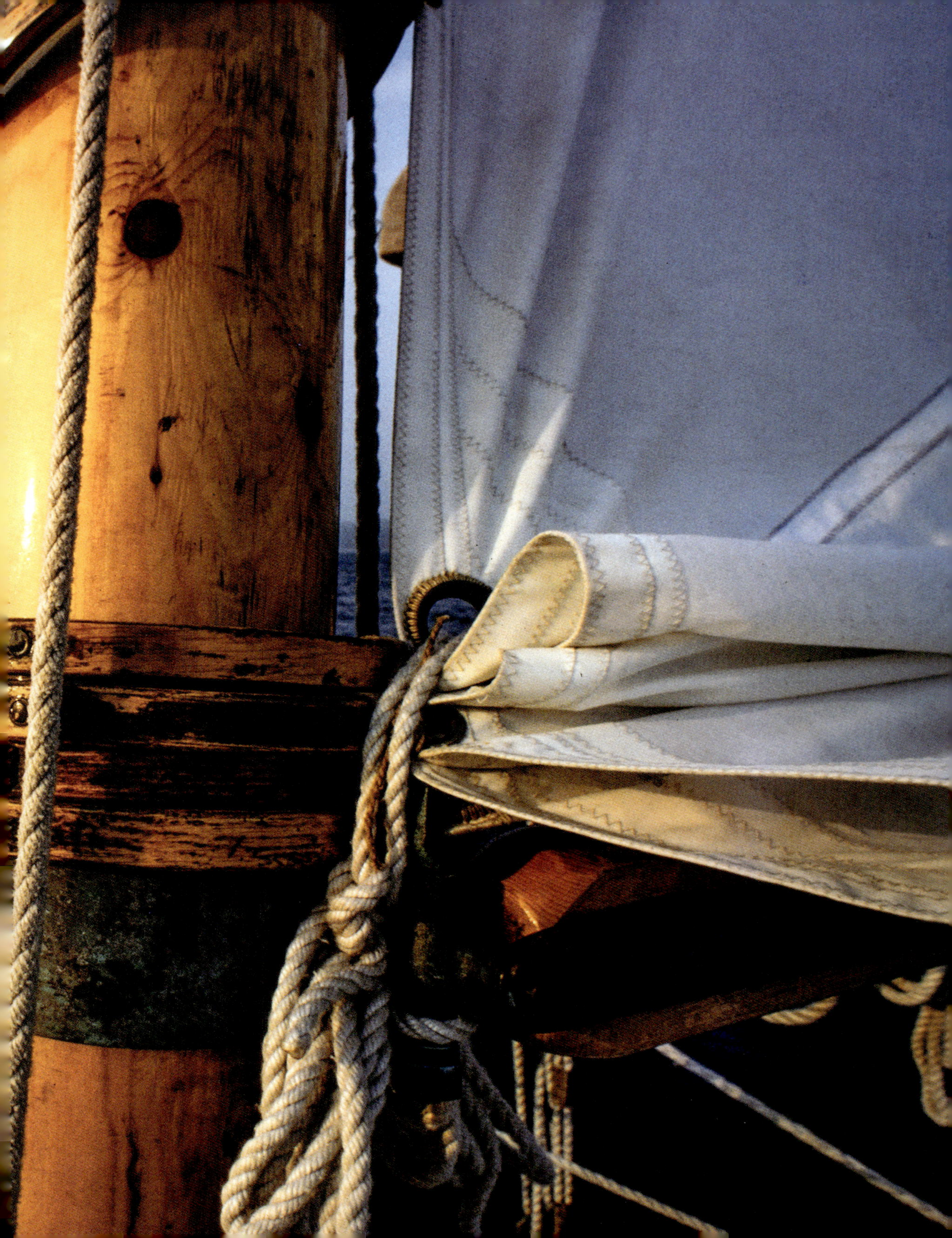

Acknowledgments

This book would not be possible if not for my friend, author, educator, and fellow photographer Lauren Walsh, who was gracious enough to introduce me to Mary Bisbee-Beek (who has since become my publicist), who introduced me to Bruce Rutledge of Chin Music Press (who is now my publisher), who in turn introduced me to Elan Robinson, who did a bang-up job on the final design of this book.

I would also like to thank Bob Shanebrook, retired Worldwide Product Manager, Professional Films at Eastman Kodak Company, who was gracious enough to proofread my manuscript to better ensure that everything you read in this book is accurate enough to take to the bank.

Let's also take a moment to thank the thousands of people who directly made Kodachrome Film images possible: the untold number of Kodak technical and businesspeople, the processing lab owners and technicians, clerks in the local photo shops, and other photographers who created the demand for the film.

And lastly, an extra special thank you to the person who puts up with me on a daily basis, my wonderful bride, Laini. (I really love ya!)

Sources

In addition to my personal experiences shooting Kodachrome transparency film professionally for thirty-plus years, this book is based on information learned from numerous places, including the many Kodak sales and technical representatives I have known over the years. Additional information and anecdotes were gleaned from the following sources.

Collins, Douglas. *The Story of Kodak*. Harry N. Abrams Publishing, 1990.

Barrett, Bill and Susan Hacker Stang. *Kodachrome: End of the Run*. Webster University Press, 2011.

Rijper, Els. *Kodachrome: The American Invention of Our World, 1939-1959*. Delano Greenidge Editions, 2002.

Dmitri, Ivan. *Kodachrome and How to Use It*. Simon & Schuster, 1940.

Wechsberg, Joseph. "Whistling in the Darkroom." *The New Yorker,* November 10, 1954.

Kessler, Herbert et al. "50 Years of Kodachrome." *Modern Photography Magazine,* September 1985.

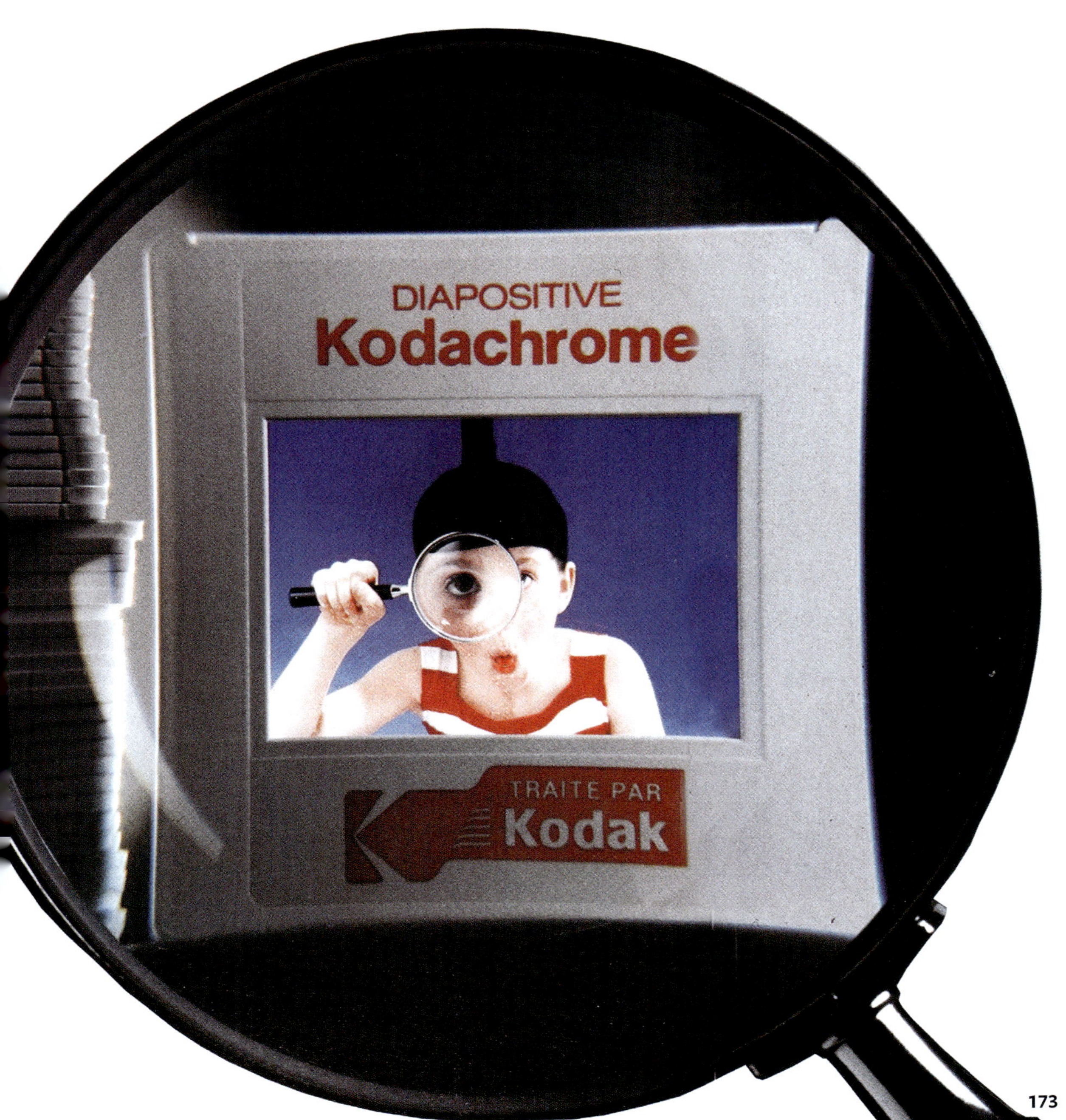
DIAPOSITIVE
Kodachrome
TRAITE PAR
Kodak

About the Author

Allan Weitz is a photographer, writer, podcast host, and educator with over five decades of experience behind a camera.

A graduate of the High School of Art & Design and the School of Visual Arts, Allan's photographs have appeared on the covers and inside spreads of major national and international publications. His photographs have won top honors from major design and advertising exhibitions. Allan has written extensively for numerous photographic publications and has served as a technical consultant for a number of photographic equipment companies. From 2015 to 2025 Allan was the host of the B&H Photography Podcast.

Additional photography by Allan Weitz can be found on his website at (allanweitz.com) Facebook (facebook.com/allan.weitz1), and Instagram (@allanweitz).

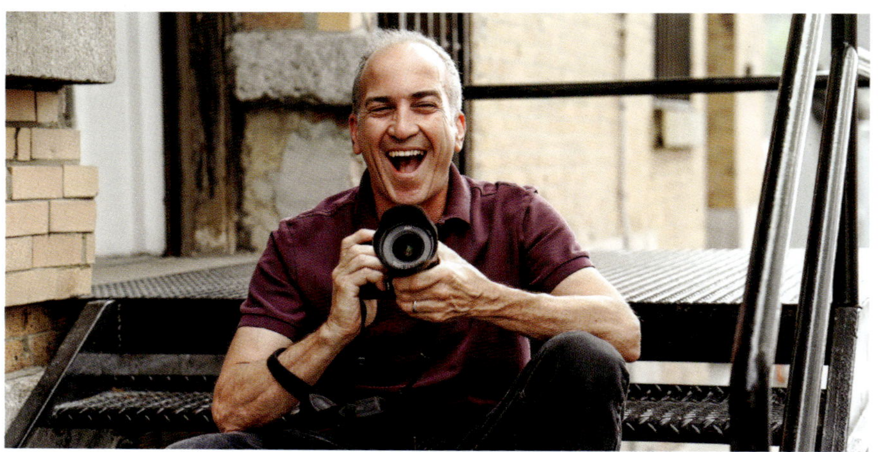